SOTHIS PRESS

Sothis Press aims to completely reimagine the public domain book with an emphasis on universal access to captivatingly designed print editions. We believe that technology will enable every single book ever published to be perpetually accessible in beautiful print form. Sothis is beauty in ubiquity.

www.sothis.press
1 Waterhouse Square, London, ECIN 2ST

Through the Yukon and Alaska

T. A. Rickard

SOTHIS PRESS

ISBN 978-1-0365-1767-0

Through the Yukon and Alaska
is in the public domain.

THE PROSPECTOR.

Through the Yukon and Alaska.

By T. A. Rickard

ASSOCIATE OF THE ROYAL SCHOOL OF MINES; EDITOR OF THE MINING AND SCIENTIFIC PRESS AND THE MINING MAGAZINE; FORMERLY STATE GEOLOGIST OF COLORADO.

SAN FRANCISCO
Mining and Scientific Press
1909

Printed by
NEAL PUBLISHING COMPANY,
San Francisco.

Dedication

To John

It was worth while to go to Nome,
To Tanana, and Nunaveck,
That I might have your welcome home
And feel your arms about my neck.

When you have grown to be a man
The world will seem a narrow strait,
From Singapore to Ketchikan
No shore will be inviolate.

But yours will be a nobler chance:
A voyageur in realms of thought,
To steer through shoals of circumstance
To isles of truth we vainly sought;

And fresh horizons you will see
Beyond the hills that shut our view
And vistas of philosophy –
Forever old, forever new.

God give you grace and make you strong
To rise above all meaner things;
And be your life as sweet a song
As that your mother sings.

T. A. R.

CONTENTS

LIST OF ILLUSTRATIONS

LIST OF ILLUSTRATIONS.

LIST OF ILLUSTRATIONS.

LIST OF ILLUSTRATIONS.

LIST OF MAPS

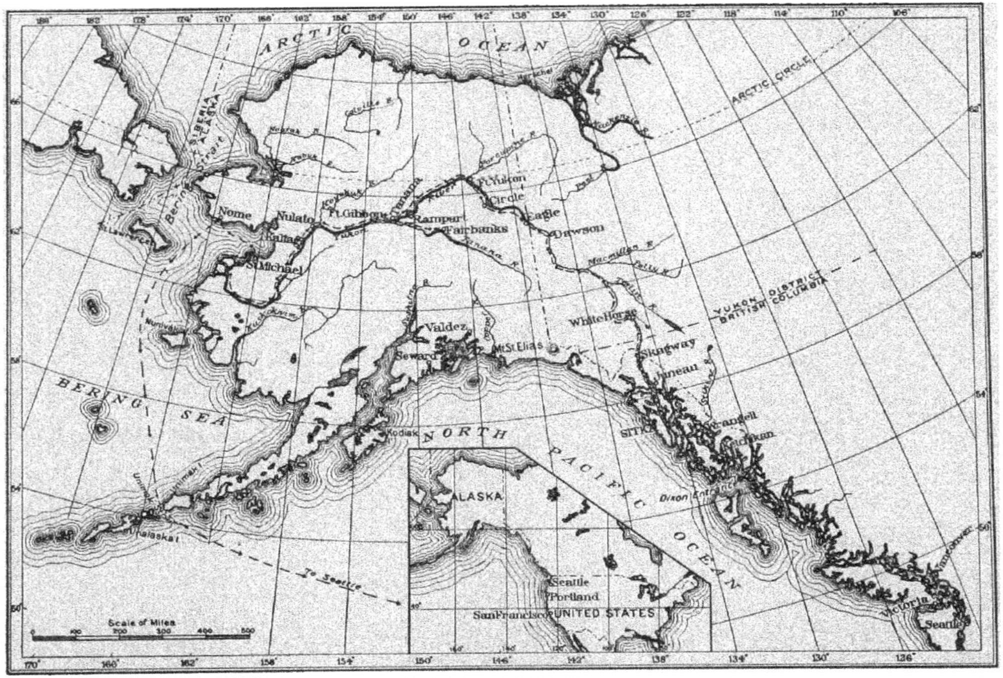

MAP OF ALASKA.
Showing the journeys recorded in this book.

Through the Yukon and Alaska.

CHAPTER I.

THE INLAND SEA.

The quiet of evening lies like a benediction on the dreaming earth; the air is still; the ship ploughs her way to the sound of many waters. Although eight bells struck an hour ago, the twilight is luminous. It is late in June; we are on the steamer *Jefferson,* bound from Seattle to Juneau. Our vessel is moving swiftly through calm waters separating the islands that fringe the coast of British Columbia and southern Alaska. It is a fiord 1000 miles long, with the blended beauty of Norway and New Zealand. On the port side the land overshadows the narrow waterway and the dark eddies are fringed with silver; in the distance the shores of an island are surmounted by high mountains flecked with snow and hooded in mist. On the starboard side the silvery waves break on a sandy beach, above which the forest extends inland to low hills, silhouetted against a gray sky, and beyond them are more dark mountains outlined against a pallid background of cloud. Ahead is a narrow space of water between two mysterious shores; we seem to be explorers plunging forward into an unknown region; we feel as if we were the first to penetrate this mysterious wonderland. That impression is constantly renewed. Narrow inlets are framed in a theatrical perspective; between ranges of hills are films of blue mist, giving the scenic effect of depth; the prospect widens continuously, one horizon succeeds another, hills upon hills arise, and there are hills upon hills behind.

No vestige of man appears; neither his work nor his dwellings are in sight. A new land untouched by the devastating hand of civilization; a land of mystery and beauty unsoiled by the wheels of industry! We pass through straits where the tide races furiously; we advance smoothly along echoing reaches separating verdant islands from wooded shores; and yet for hours we see no trace of human handiwork. We might be with Vancouver feeling his way amid uncharted seas; we might be with an expedition sent from Sitka by Baranoff in search of furs; we might be the companions of a Hudson's Bay trader poaching on the preserves of the Russian fur company. Of later British energy there is no sign; of the latest American exploitation there is no suggestion. We have gone back at least a century and are cruising along shores never trodden by the men of our race!

Evidence of modern industry is not wholly lacking, as is discovered later. A saw-mill or a pulp-mill is detected on the edge of a bay, and behind it the flash of a waterfall suggests a source of power for machinery. Salmon canneries remind the traveler that the fisheries of Alaska are gold mines. A smelter, a tramway up the steep hillside, a group of cabins in the forest, a white scar on the mountain—all these suggest the activities of the miner. But such evidences of industry, while collectively important, separately are insignificant in the vastness of the region. Pictorially, they play a part only at rare intervals.

The shores of the islands, and of the mainland, are thickly wooded. Spruce and hemlock prevail, but the trees are small, and unsuited for lumber. In an acre of forest only three or four spruce will exceed 5 feet in diameter. The hemlock, when

THE MIDNIGHT SUN.

thick enough, is used in the mines. At an altitude of 500 to 1000 feet above tide-water yellow cedar succeeds the spruce, and at 1000 to 5000 feet only scrub pine or piñon and other low brush finds a foothold. The scars made by landslips and snow-slides are healed by a thick growth of alder, in patches of velvety green. The soil is scant, the forest growing from a tangle of decayed vegetation and moss. When the large trees are cut down, this moss dries, and the forest is killed. Prospecting is difficult, for the explorer has almost to hew his way. The perpetually misty climate of this southeastern portion of Alaska favors a growth tropical in its luxuriance. During the short summer of four or five months the vegetation waxes riotous. While walking from Juneau to Silver Bow basin during June I noted that the roadside was already thickly fringed with white spirea, red columbine, and pink huckleberry; all the shrubs were ready to burst into bud while still under their coverlet of snow, flowering before the leaves were out. At Treadwell the violets had a longer stem and a lovelier color than the Neapolitan, and with them went a perfume exquisitely delicate.

The international boundary between British Columbia and Alaska runs through Cape Chacon, the southern extremity of Prince of Wales island, and thence northeastward up the narrow fiord called the Portland Canal. At Cape Chacon, Juan Perez, the Spanish navigator, landed in 1774; finding a native with a Russian gun in his hands, he marked the line of 54° 40′ north latitude as the limit of Russian rule southward and of Spanish dominion northward. This proved strangely prophetic as to the boundary, but an ironical fate decreed that neither the Russian nor the Spaniard should long enjoy it as a line of international contact.

"Fifty-four forty" became the war-cry of a belligerent party led by Lewis Cass in 1843, when Great Britain and the United States were quarreling over the international boundary. "Fifty-four forty, or fight!" was the motto of the supporters of Polk, but in 1846 Buchanan signed the Oregon treaty making the forty-ninth parallel the line of demarcation between American and British territory in the Northwest.

The *Jefferson* is 216 feet long and is rated at 1615 tons. She flies the flag of the Alaska Steamship Co. Everything aboard is scrupulously clean and the food is good. Nevertheless, I venture a criticism: dinner is served at 5 p.m.; owing to the long daylight the time seems like mid-afternoon, and to many

THE 'JEFFERSON' IN QUEEN CHARLOTTE SOUND.

of the passengers the custom is annoying. Another criticism can be made upon the elaborate character of the *menu*; it is absurd in the small galley of a ship like this to attempt to give a dinner that if properly done would tax the resources of a large kitchen. The consequence is number without variety, for the various dishes are cooked so closely together as to lose their distinctive flavor. There may be persons who travel on coastwise vessels to enjoy French cooking, and there may be those that go away from home to dine better, but is it not unfair to make the average traveler suffer for the sake of such simpletons? While dwelling on these minor matters, it is worth recording that on none of the dozen steamboats on which I traveled in this journey to the North did I find a bath or a bath-room available. There was a bath on one boat, but it was used for pickling pigs' feet; there was a bath-room on another steamer, but it was filled with stores. If cleanliness be next to godliness, you do not travel heavenward when you go northward.

The dark woods are turned gray by dead trees, which stand out amid the green like the gray hairs of a badger. This is due to the decay of trees that have reached the maximum growth permitted by the thin soil and short summer. Most of this timber is only fit for making pulp to be used in manufacturing paper. It is a curious example of national lack of thrift that the lordly forests of California and Oregon should be destroyed for making paper-pulp when these Alaskan woods, useless for any other purpose, are so readily available. The climate being excessively humid, no harm would be done in this case by the destruction of the forest, viewed as a preserver of moisture. Moreover, the soil on the islands being non-productive, the destruction of the trees, and later of the moss, would facilitate exploration and expedite the search for mineral wealth.

Ketchikan is a typical Alaskan port. It is the distributing point for the miners and fishermen on the islands of Revillagigedo and Prince of Wales. The town is on stilts, that is, it is built on piles; over them are laid the weatherbeaten gray boards that constitute streets; these are scrupulously clean, for no horses traverse them. We happened to land on a Sunday morning and as we strolled along the quiet avenues, the strains of a hymn came sweetly from a Methodist church. Ketchikan wears a sober look; it is far from the stamping ground of the great herd; in whatever rampaging its own inhabitants might indulge, such exuberance would raise no dust, for a mist is in the air and water is underfoot.

At 9 o'clock on June 21 we are threading the famous Wrangell Narrows. This is a strait 19 miles long, tortuous, and in places only 100 yards wide. It was long considered navigable only by vessels of the lightest draught, until Capt. J. B. Coghlan, U.S.N., surveyed and buoyed the channel in 1884. At low tide there is a bar that leaves only 14 feet of water, but as the tide rises 10 feet there is ample clearance at high water. The *Jefferson* drew 15 feet when we left Seattle; her commander is himself the pilot. In the fading daylight

KETCHIKAN.

our course is steered by aid of white monuments of concrete surmounted with white crosses on shore, while in the water the red buoys guide the navigator in finding the narrow channel in a strait that is so serpentine and shallow as to seem more like a river than an arm of the sea. On one small island we see a fence enclosing a grave that marks the last prospect hole of a pioneer, who was killed by the Indians, and just beyond a half-sunken barge lies wrecked on a reef. The wild duck examine it curiously. On several trees eagles are perched; on others, the grouse; while overhead a flock of geese flies athwart the sky in characteristic procession. It is said that on the islands deer abound; but on the mainland they are scarce, because the timber wolves hunt them.

The daylight fades, but we can see the green moss along the shore and the russet seaweed floating on the edge of the waterway. A bend brings to view a deserted saltery, for it is too early for the fishing season; another turn in the channel and a small settlement emerges from the forest. The blue smoke curls peacefully in token of rest after labor. We issue from the narrow strait into a mystic lake. The succession of pictures is so rapid that nothing can be entirely unexpected and the attention is held so strongly that even at midnight in this luminous atmosphere the senses are wakeful and sleep seems but waste of time.

Entering Lynn Canal, the fiord is six miles wide. On a gray day the scenery is grandly desolate. To the west is a mountainous coast; being on the cold side of the range, the snow-fields reach to a broad band of dark forest, which, in purple shadow, mantles the foot of the mountains to the water's edge. Torn fragments of mist fly wildly and the wind blowing off the ice-fields is as cold as man's ingratitude. On the eastern side the mountains rise abruptly for 5000 feet above the tide, and the timber-line is clearly marked at 1500 feet. Above this level the bare rock, in brown and purple, with patches of green moss, reaches to the snow-fields. In every ravine a cascade comes tumbling in reckless haste and on the crest the piled mass of *névé* marks the foot of the glacier. One of these perched high on the rock-slope seems ready to fall upon the ship. The sun-light breaking through the clouds irradiates the blue cliffs of ice and places a coronet of sapphire upon the mountain's brow. To starboard, to port, aft, ahead, wherever the eye turns, are snowy mountains, blue ice-fields, and gray skies. We are entering the Northland.

CHAPTER II.

DISCOVERY AND DEVELOPMENT.

Alaska is a great land. That is what the word *alaksa* originally signified. When the first Russian adventurers reached

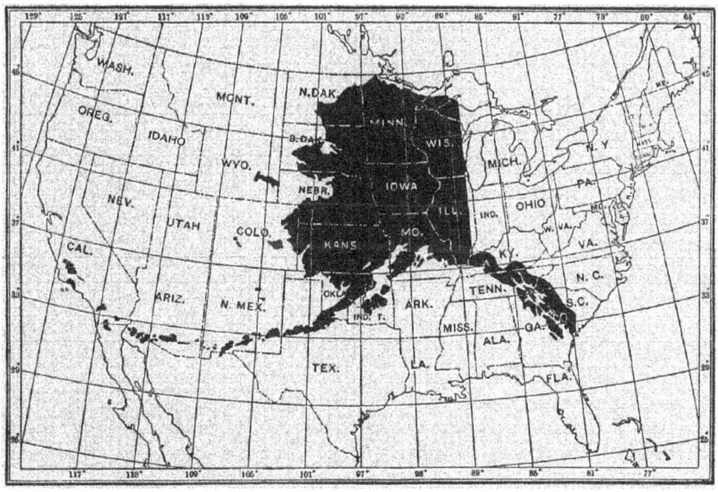

MAP SHOWING RELATIVE SIZE OF ALASKA.
(After U. S. Geological Survey.)

the Aleutian islands they were told by the natives that eastward lay a great stretch of country, which they called *al-ak-shak* or *al-ay-ek-sa*. The island of Unalaska was then known among the natives as *Na-gun-alayeksa*, or "the land near Alayeksa." In time the native name was corrupted to Alaska; it is an English version, for the Russians never used it. Thus the name was in effect a prophecy, the true significance of which was not understood until 240 years after the first European landed on the northern coast of western Alaska.

The relative size of Alaska is shown by the accompanying map. This illustrates the fact that the east and west points of Alaska are as far apart as the Atlantic is from the Pacific in the latitude of Los Angeles, while the northerly and southerly extremities are as widely separated as the Mexican and Canadian boundaries of the United States. It is often said that San Francisco is east of the centre of the United States. This apparent paradox is explained by the fact that the most westerly island of the Aleutian chain is farther west from San Francisco than San Francisco is from New York. The Seward Peninsula at Cape Prince of Wales is only 60 miles from the Siberian coast of Asia. The distance by sea from San Francisco to Skagway is 1696 miles; from Skagway to Nome, by the Yukon, is 2274; and from San Francisco to Nome, 2731. Owing to the fact that Alaska is usually shown on maps either by itself or as part of the continent of North America, most persons acquire wrong conceptions of its size and position; the study of maps is a remedy, and a journey to Alaska a cure, for any such misunderstanding.

The early history of a country is linked to its topographic features. Mountains are barriers, rivers are avenues. The first foreigners to greet the natives of this big corner of the American continent came from Asia, for the Pacific afforded an approach to the islands that, like sentinels, are thrown far out to sea from an inhospitable shore. The Russians crossed Siberia and explored the Arctic coast of Asia. In 1728 a band of Cossacks was driven by a storm eastward, landing in Norton Sound. Others came across from Kamchatka and settled on the islands and peninsulas of southeastern Alaska. The mountains guarding the coast discouraged exploration into the interior. Another range—the extension of the Rocky Mountains—barred the westward progress of the French *voyageurs* and the English fur-traders of the Hudson's Bay Company. After the Russians had obtained a foothold among the Indians, the vice-

ON THE TURN OF THE TIDE.

roys of Mexico sent successive expeditions up the coast, such as that of Perez in 1774. If the English fur-traders had not interfered, the Spanish and Russian spheres of influence would have conflicted and a contest for control would probably have ended in the establishment of the Sacramento river as the line of demarcation. The British navigator Captain Cook landed near Sitka in 1778, while seeking a way by water to Hudson's Bay. Fifteen years later his midshipman, Vancouver, surveyed the coast carefully and completely from latitude 35 to 60° north. Meanwhile the English were finding a way overland from Canada. Mackenzie, in behalf of the Northwestern Fur Company, ascended the Peace river, crossed the Rocky Mountains, and reached Pacific tide-water in Queen Charlotte sound, only to learn that Vancouver had preceded him by a short interval, in 1793. Thenceforward the known portion of Alaska, from Unalaska along the fringe of islands to Sitka and thence to British Columbia, was the battle-ground between the agents of the two fur companies, namely, the Russian American Company and the Hudson's Bay Company. Not until 1826 did the Russians extend their explorations along the northwestern coast to the mouth of the Yukon. The establishment of a post at St. Michael prepared the way for trade up the great river of Alaska. In 1843 Zagoskin reached the mouth of the Tanana and built Nulato.

While the Russians were exploring the west coast of Alaska, the English were finding their way along the Arctic. In 1789 Mackenzie descended the river that now bears his name and reached the frozen sea. In 1826 Franklin went westward from the mouth of the Mackenzie. Then the relief expeditions sent from England (between 1845 and 1853) in search of Franklin explored and charted portions of the Seward Peninsula.

The great interior region was still unknown, although the Hudson's Bay Company was persistently advancing its outposts westward. In 1840 a 'factor' or agent of that company established a trading post at the head of the Pelly, a tributary of the Yukon. In 1847 Fort Yukon was built by Murray. The English traders heard that the Russians were in the lower Yukon, and in 1850 they descended to Nulato. Thus here and there at enormous distances apart the lonely outposts of the European races were gaining a foothold. The only object of their intrusion into the inhospitable wilderness was the trade in furs. No whisper of gold was heard.

In 1863 the Western Union Telegraph Company sent an expedition to survey a telegraph line that was to connect America and Europe, by way of Asia. Submarine transmission by cable under the Atlantic was believed to be imprac-

ON THE MUSH.

ticable. The survey of the proposed route through British Columbia, Alaska, and Siberia, involved the exploration of regions but little known. In Siberia, George Kennan did good work; in Alaska, Robert Kennicott was the leading spirit. Although the project of a telegraphic system was nipped in 1867 by the announcement that the Atlantic cable was a success, the explorations made then and thereafter by the men in charge of the Western Union expeditions proved most important. They ascended the Yukon, and they crossed the Seward Peninsula. The information they procured proved of great value in the negotiations between the American and Russian governments at the time of the transfer, and the routes they mapped were followed by the telegraph lines built as soon as the country became famous for its gold deposits.

In 1867 Russian America was purchased by the United States for the sum of $7,200,000, and the 'district,' at the suggestion of William Seward, the Secretary of State, was named 'Alaska'. At that time the finding of gold had been reported in a vague way, but no profitable mining had been done. The Chilkoot Indians opposed the incoming prospector until 1880, when 16 miners, under Edmund Bean, crossed the Chilkoot pass and descended the upper branches of the Yukon. In 1883 Frederick Schwatka crossed the same pass and followed the Yukon all the way to the sea. His graphic account of the expedition appeared in the *Century* magazine and did much to excite interest in Alaska. But an event even more important was the voting by the American Congress in 1895 of a small appropriation enabling the Geological Survey to send a party into Alaska. From that year up to the present, successive parties of scientific explorers have carried the investigations of the Survey across the rivers, mountains, and morasses of Alaska, doing a work the value of which is now fully appreciated.

A little desultory gold and copper mining had been done in a few localities, but with the exception of the great Treadwell mine and one or two others in the vicinity of Juneau, mining in Alaska was a negligible quantity. Then suddenly, as out of a clear sky, came the tremendous shout of a big boom, with all the excitement that follows wonderful discoveries of gold. In 1896 George Carmack found gold on the Klondike, in the Yukon Territory. A mob of 50,000 adventurers rushed to the diggings. In 1898 the golden beach of Nome was discovered and another stampede ensued. Alaska had arrived.

CHAPTER III.

JUNEAU.

On the morning of the fourth day from Seattle the *Jefferson* reaches Juneau. It it early dawn; the mists are climbing the wooded slopes of the mountains that border the straight course of Gastineau Channel. To our left, or westward, the dwellings, offices, and shaft-houses of the Treadwell mines form a long settlement along the shore of Douglas island, whose higher contours are surmounted by the snowclad peak named Jumbo. Close to the water are several large buildings emitting the muffled roar that proves them to be stamp-mills. The red head-frame of No. 2 shaft of the Alaska Treadwell mine is silhouetted against the gray wall of the cavernous opening called the 'glory hole'. On the beach a gray building, resembling a natatorium, is the club where the miners congregate. The big bunk-houses, one of them in process of repair, suggest other human aspects of the mining business. On the first rise above the shore are a number of new cottages, giving a touch of the picturesque to this industrial settlement. A long wharf indicates the magnitude of the trade in supplies and machinery arising from the operation of mines producing $3,250,000 per annum and employing 1200 men. Oil tanks, freshly painted bright red, punctuate the foreshore and assert the economy of liquid fuel over coal. They give a chromatic liveliness to the quiet landscape. Behind the wharf the residence of the general manager suggests the watchful skill dominating large operations, while the lace curtains and neat lawn bespeak the womanly grace that makes of every abiding place a home.

Two miles northward, up the channel, on the other side, which is the mainland, the pretty town of Juneau lies ensconced in the lap of the mountains guarding the passes into the Northland. Above the town is the alluvial fan at the mouth of Gold creek, the stream that led the pioneers of 1880 to the rich deposits of Silver Bow basin, a glacial cirque, five miles above Juneau.

It has been said that Juneau is the gateway to the mining regions of Alaska; undoubtedly, this pioneer settlement has been the point of departure for the adventurous spirits that explored the wilderness and laid the foundations of existing industrial development. Up to the time of the transfer of Alaska from Russia to the United States, in 1867, there had been no gold mining. The Russian governors, of whom Baranoff was chief, had discouraged the search for gold because it might have interfered with the fur trade, which was their source of profit. For gold mining they had no liking, and of it they had no knowledge. Some old records prove that the Russians had observed the occurrence of gold in several localities but made no effort to exploit the deposits. The Stikine river, about 1865, was invaded by prospectors and in 1874 the Cassiar diggings were established on the Canadian side of the boundary, just out of Alaska. In 1869 some of the miners from the Stikine went north and made placer discoveries on Windham bay and Sumdum bay. In 1870-'71 about $40,000 was obtained from these two localities. This was the beginning of gold mining in Alaska.

Near Sitka, on Baranoff island, mining began in 1877, with the location of the Lucky Chance and Stewart on Silver bay. One of the operators on the Stewart was George E. Pilz, who erected the first 10-stamp mill in Alaska on that mine in 1879. Rumors of gold had been afloat among the people of Sitka for many years before, and ever since. It is said that more than 30 years ago the Auk Indians used to come to Sitka wearing gold ornaments; for Sitka was then the centre of the coastal trade. In 1878 John Muir was deputed by the United States Government to explore southeastern Alaska, and in his report he stated that the region between Windham bay on Stephens passage, about 65 miles southeast from the site of Juneau, and Sullivan island, 60 miles northwest, in Lynn Canal, would make

MOONLIGHT ON SUMDUM BAY.

a second California. This report in pamphlet form came to Sitka in the early spring of 1880. George E. Pilz had just completed his stamp-mill. He and N. A. Fuller, a local merchant, after reading Muir's report, decided to send prospectors into the gold-bearing region described by the celebrated naturalist. As soon as spring opened and the snow was off the mountains, Pilz and Fuller engaged Joseph Juneau, a French Canadian, and Richard T. Harris, an American. These two miners were properly outfitted; accompanied by three Indians, they started on July 19, 1880. From Sitka they went in a small boat first to Windham bay, then to Sumdum, then back to Windham bay, where they located several claims. Crossing Taku inlet, a treacherous passage, they sailed north to the head of Lynn Canal. Coming back they called at the Auk village 12 miles north of the site of Juneau and obtained some information from the Indians. Skirting the eastern shore they ran aground on the bar of Gastineau Channel at the north end of Douglas island, and camped on August 16. While prospecting they found a creek so full of dead salmon that they named it Salmon creek, the name it now bears. Rowing four or five miles farther south along the east shore they came upon another stream, at the mouth of which they found sand containing gold. This they named Gold creek. The date was August 17, 1880. They made their way up the canyon for about two miles, where they found some rich quartz veins and located several claims. Being short of provisions, the two explorers went back to Sitka on August 23, but returned immediately to Gold creek, which they examined carefully from its mouth to Silver Bow basin, a distance of five miles. Many more claims were located. With the aid of the Indians they cut a trail and packed 800 pounds of specimen ore, with which they returned to Sitka in November.

Juneau and Harris gave a frank account of their discoveries. Inevitably, there was much excitement in the little frontier outpost. A stampede followed. Among the first to go was Edmund Bean, who camped on the site of Juneau, then covered with a forest of spruce and hemlock. Juneau and Harris, together with five others, hurried thither in a steam launch

THE SUMDUM CHIEF MINE.
(Photographed by Moonlight.)

borrowed from the U. S. gunboat *Jamestown,* then lying in the harbor of Sitka. Eleven other men, including John Olds, Hugh Campbell, and John Dix, hired the steamer *Favorite* and left Sitka on November 26. On arrival they found that the preceding party had located by proxy, which was contrary to the regulations of the Cassiar district, where most of the prospectors had obtained their notions of mining law. The townsite also was located, covering everything except what is now Main street. During the winter a dozen cabins were built. In the spring of 1881 the Northwest Trading Co. sent a representative, Edward De Groff, to open a store. He came in a sloop with a party of seven others and a stock of supplies. Shortly afterward a post-office was established with De Groff as postmaster.

In October 1880 before returning to Sitka, Dick Harris and Juneau, with three natives, held a so-called miners' meeting on the site of Juneau. Harris wrote the laws of the mining district. The preamble states that the meeting was organized "by Richard T. Harris, Joseph Juneau, and three representatives of Geo. E. Pilz and N. A. Fuller." The natives and Juneau could neither read nor write, so that everything was left to Dick, who named the future town Harrisburg and also gave his own name to the mining district.

Harris and his friends returned to Sitka, bringing the books with them; when the rush commenced and the miners arrived at Juneau, they found no way to record their claims, the Recorder and the books being at Sitka. A strong demand was made for the return of both. Harris sent the records in the *Jamestown* launch in February 1881, by the hands of Lieut. Commander C. H. Rockwell. At a meeting called soon after, Richard Dixon was elected Recorder, and Harris being unpopular just then, the 72 miners present voted to christen the town Rockwell, to which Rockwell objected; therefore, the name Harrisburg was retained until December 1881, when at a mass meeting called to settle disputes and make town regulations, the name of the settlement was changed to Juneau. Harrisburg was "too commonplace." De Groff was secretary of the meeting and the minutes now form part of the records at Juneau.

Some of the resolutions were unique in their way but they have been used as precedents in recent litigation. Poor Dick Harris had tried hard to perpetuate his name but he was destined to lose the distinction. First the name of the town was changed to Juneau, as described above, and when the District of Alaska was organized in 1900, the Harris Mining District became the

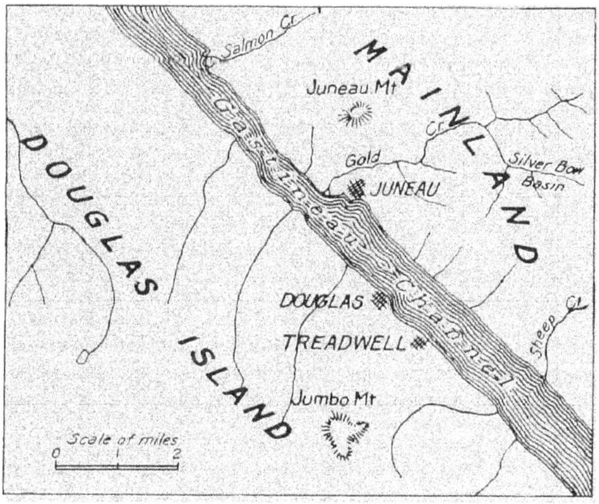

JUNEAU AND VICINITY.

Juneau Recording District. Thus Dick's illiterate partner won fame without an effort.

John Olds, now proprietor of the Occidental hotel, at Juneau, was the second man to reach (in April 1881) the ridge above Silver Bow basin, where the outcrops of the Alaska-Perseverance and Groundhog lodes cross the mountain. He tells me that wherever the snow was gone he saw lots of loose 'float'—pieces of ore from a vein—and in these fragments of quartz he could see gold. The veteran acknowledges that he became excited and thought he had found "the richest country on earth." He located the claims, but when the experts came and condemned the discovery, he dropped his locations. Subsequently the 'float' was gathered and carried for treatment to a little *arrastre* that was built near the head of the present Perseverance ditch. Another of the pioneers of Alaska is Hi Chung or 'China Joe', who came to Wrangell in '74 and to Juneau in '81. When the Chinese were driven out in 1885, he was allowed to remain, because he had a house and lot, and a bakery. "Many a hungry white man and Indian has he fed, you bet," and the prospectors are always owing him money for bread. He attends meetings of the pioneers and is proud to participate.

Juneau was incorporated on June 29, 1900, and in 1906 it became the capital of the District of Alaska. Thereby hangs a tale. At the time of the transfer, in 1867, Sitka was chosen as the headquarters of the commanding officer attached to the department of the Columbia; in 1879 a naval officer was placed in command; in 1884 a civil government was established, consisting of a governor, a judge, a marshall (who was ex-officio surveyor general), a district attorney, and a clerk of the Court. At first there was only one judicial district, but later two more were established. Until 1900 the laws of Oregon were made applicable to Alaska. In 1900 a code of laws was given by the Congress of the United States and in some of these enactments the District of Alaska is first called a 'territory'. In 1906 Alaska was given the right to elect a delegate to the Congress, like the other Territories. The code of 1900, known as the Carter code, established the seat of government at Juneau, providing however that it should remain at Sitka until suitable buildings became available "by purchase or otherwise" at Juneau. In May 1906 the Congress arranged that the contingent fund appropriated for the Governor of Alaska should be available for the rent of an office and residence at Juneau. Suitable buildings were leased and the seat of government was transferred from Sitka to Juneau on October 1, 1906, by Wilford B. Hoggatt, the sixth Governor of Alaska.

CHAPTER IV.

THE TREADWELL MINES.

The beginnings of big enterprises are always romantic when viewed through the telescope of success; the story of a mine that has yielded millions in gold will command interest, especially among those engaged in the search for a duplicate. Herewith is the tale of the discovery and development of the Alaska Treadwell, which has produced $22,500,000 and paid $10,500,000 in dividends:

The man who found the Treadwell lode was Pierre Erussard, a French Canadian, known to his acquaintances as French Pete. He lived among the Indians and was a prospector. Some of the veterans describe him as tall, well built, and dark, with the black hair, mustache, and tufted beard common to his countrymen. When Juneau and Harris came to Sitka with their news concerning Gold creek, Pierre started forth on a similar quest, accompanied by several Indians, one of whom was his wife's brother. They landed on the beach of Douglas island in November 1880 and found gold in the sand. Pierre also found an outcrop of gold-bearing quartz on the hillslope, about a quarter of a mile from the shore. He located two claims to cover this ore; one of them was called the Paris, after the capital of France, and the other he named the Bear's Nest, because he found the ground occupied by a bear and her two cubs.

The original discovery was made on the west side of Paris creek, a streamlet, long since obliterated, that formerly ran down the slope now deeply scarred by the big excavations of the Alaska Treadwell mine. The creek exposed the outcrop of white quartz; but this was only a small part of the top of the lode, the rest of it being covered with blue clay containing numerous barnacles. Remnants of this deposit are still visible nearer the beach. This is evidence of a former shore-line and one proof, out of many, that southeastern Alaska is undergoing slow elevation. The disintegrated quartz of the outcrop was shoveled by Pierre into his sluice-boxes and a little later he drove a shallow adit or 'tunnel' to cut the lode a few feet below the moss-covered surface. This was the first mining on Douglas island.

The Paris claim covered the site of the present 'glory hole.' Pierre worked the small placer formed by the concentration of the gold eroded from the big outcrop of the quartz lode and he also dug into the softer superficial portion of the lode itself, washing the gold-bearing material thus obtained. He had rockers and sluice-boxes of the conventional kind, with which he did fairly well. But it was no bonanza, compared to Gold creek. Therefore it is not surprising that Pierre sold the better of his two claims a year later. The record states: "September 13, 1881. Transfer of Paris lode from Pierre Joseph Erussard (or 'French Pete'), original locator, to John Treadwell, in consideration of the sum of five dollars ($5.00)."

John Treadwell was a builder and contractor, with experience in mining, for as early as 1869 he had worked in White Pine county, Nevada, and for 12 years before going to Alaska he had been engaged both in quartz and hydraulic mining in Nevada and California. In 1881 he had charge of the building of a house for John D. Fry, a banker who took a prominent part in the early development of San Francisco and was one of the founders of the California Safe Deposit & Trust Co. To Colonel Fry and to his friend James Freeborn there came the story of a rich prospect in the hills behind Juneau. Looking around them for a man to be sent in their behalf to inspect the mine, Fry suggested Treadwell as being both trustworthy and possessed of enough mining experience to be able to appraise a prospect. It was arranged to pay Treadwell's expenses and to give him a one-third interest in the mine, if the purchase was recommended. Treadwell went north, saw the prospect, and found that it was a stringer of quartz carrying free gold. This

LOOKING ACROSS GASTINEAU CHANNEL, FROM JUNEAU TO TREADWELL.

Photograph by Winter & Pond, Juneau. Published by permission.

small vein was in the slate belt close to the present Ebner mine, near Silver Bow basin, and whatever other unfavorable marks it may have had it was plain to Treadwell that most of the rich ore had been dug out. Naturally disappointed, he decided to return to San Francisco, and it was while waiting for a steamer at Juneau that he first met French Pete (otherwise Pierre Erussard), who had opened a store in that town. Pierre happened to need ready money to pay for freight on stores that had just arrived from the south; he wanted $500, and was willing to accept that sum for an interest in his mine on Douglas island. Without going to see the claim, Treadwell "took a flyer," and advanced $500 from the funds intended for the purchase of the other mine. Then he went across the water to see what Pierre possessed. Treadwell liked the look of the lode and took a bond on the Paris claim for $20,000. He then went prospecting, spending the remainder of the time before the close of the season in testing his new acquisition. Before returning to San Francisco he stayed with Pierre for two weeks. Treadwell stated that the Paris ore was too low-grade and suggested that if Pierre would give him a quit-claim deed for $5 he would try to sell the mine in San Francisco, and would undertake to trade at Pierre's store if the sale were effected. The deal was made. Treadwell went to San Francisco and returned on May 17, 1882, with a 5-stamp mill, which he erected on the Paris claim. Fry and Freeborn completed their agreement, and thus Treadwell got a third of the mine. A few years later Freeborn was prompted by ill health to sell his interest, which was then offered to D. O. Mills, who, after making a trip to Alaska personally to inspect the mine, decided to make the purchase, and thus obtained Freeborn's holding. The Alaska Mill & Mining Co. was formed and controlled the Paris mine, together with adjoining property, until June 1, 1890, when the Alaska Treadwell Gold Mining Co. was incorporated under the laws of Minnesota. In deference to his great business ability, Mr. Mills has always remained in control, though a majority interest was purchased in 1890 by the Exploration Company, of London, on the advice of Hamilton Smith. He was the first consulting engineer to the Alaska

THE GLORY HOLE.

WORKING IN THE GLORY HOLE.

Treadwell Co. and was followed successively by H. C. Perkins, Thomas Mein, and F. W. Bradley. Thus this great mining enterprise was started. In 1883 work was begun on the first large mill, of 120 stamps, and was completed two years later. In 1887, 120 more stamps were placed under the same roof. Between 1893 and 1899 the Mexican, Seven Hundred Foot, and Ready Bullion mills were erected, and the new 300-stamp mill of the Alaska Treadwell. Thus 880 stamps were put to work. All of these are now active, treating 1,360,000 tons and yielding $3,250,000 per year.

Every visitor to Douglas island climbs the short slope that leads to the 'glory hole.' This is an enormous pit made in the course of mining. The lode has been removed to a maximum width of 420 feet and for a length of 1400 feet. The deepest point of the bottom is now at 580 feet below the surface. Standing on the edge of this cavernous excavation the traveler will realize what a vast amount of ore is crushed within the relentless maw of the big stamp-mills. The 540 stamps of the Alaska Treadwell require 2750 to 3000 tons per day to keep going. From the glory hole 5,086,500 tons has been taken out. In 1895, the first underground stoping was done, but since 1905 practically all ore milled has come from underground stopes, which, up to May 31, 1908, have yielded 4,141,682 tons on a total amount milled to that date of 9,228,182 tons of ore. Even a careless observer will note that the rock is not all quartz. On the foot-wall or western face the black slate is exposed and on the opposite side, called the hanging wall, the gray-green of gabbro is contrasted with the white ore. In the midst of the ore a tongue of slate protrudes, widening to the south so as to split the lode. This will suggest something of the geology.

The orebodies of the Treadwell group of mines consist of dikes of diorite penetrating the contact between an older upturned bed of this green gabbro and the slate itself. The diorite is of volcanic origin and came from below through fractures in the crust of the earth after the manner of water rising in the cracks formed in overlying ice. Subsequently the molten rock cooled, shrank, cracked, and was penetrated by thermal waters, such as usually mark the quiescent stages of volcanic

UNDERGROUND IN THE ALASKA TREADWELL MINE.

activity. These waters contained the gold, derived from other rocks below, and as they circulated along the lines of least resistance established first by the intrusion and then by the cooling of the diorite they precipitated the gold through interstices in the diorite, together with quartz. Thus the lode was formed.

This, however, is not the place for scientific detail; I realize that my readers are with me on a holiday journey to the North and must not be bored with too much geology. Attention easily wanders from scientific considerations to others not less interesting. Looking down, the mine workings that pierce the sides of the immense excavation and penetrate into mysterious inner chambers appear like the tunnelings of a mole. A ladder resting near the bottom emphasizes the dimensions and the throwing of a stone across the void suggests the deceptive largeness of it. The accompanying photograph does not do justice to the subject because the colors are lacking: the white quartz, the blue-black slate, the fringe of green bushes, the gray stems of the spruce, and the reddish splashes where wind and weather have decomposed the iron minerals. The yellow sunshine bathes one side of the pit while deep blue shadows lurk on the other side; over the edge is the dark red head-frame of a shaft-house; beyond it the flash of waters marks Gastineau Channel, with a green shore on the farther side; and more distant still are the blue hills silver-crested with eternal snow.

But to see the real Alaska Treadwell mine you must go underground, descending one of the shafts in a 'cage,' only slightly less pretentious than an office elevator. This brings you to a 'station,' from which galleries extend into the heart of the rock. Following the rails of the car-track, you step to one side as a train of cars, pulled by a horse, comes rattling past; then you ascend a short ladder and reach a cavernous opening, dimly illumined by candles placed at the points where men are at work. The great opening is 180 feet wide and 100 feet long, it is separated by pillars 18 to 25 feet thick from similar chambers, so that a space 410 feet long and from 150 to 200 feet wide has been excavated. The gold-bearing rock

A BIG STOPE.

has been broken for a height of 160 feet above the tenth level, but as yet only a little of the broken ore has been removed, enabling the miners to stand on it while they attack the ground above them. The general effect is that of a flatly arched low-roofed cavern with an irregular rugged floor of broken ore. The lights show dimly and the chugging of the machine-drills fills the vault with sound. This stope is so large that each foot of additional height over the entire area yields 1000 tons of ore. As seen from the level the men in the distant corners of the stope look like gnomes, but the candle-light throws shadows of giant size. This stope communicates with others, both above and below.

The non-technical visitor will wonder how such openings are made without danger of the rock collapsing on the workers. The solution of this problem involves an important phase of mining engineering. In this case safety is secured by leaving a thickness of 25 feet of rock across the width of the lode at intervals of 70 to 80 feet. This makes a 'pillar' able to sustain the 'walls' and from these 'pillars' the roof arches flatly in such fashion as to support the overhanging mass of rock. Yet many men are killed. When a fatal accident happens it is usually due to carelessness on their part; experienced men can tell whether a crack means loose rock, by sounding with a hammer; a fall of rock is heralded by crackling, even for a week beforehand, and thus gives warning. But some men will take needless risks; a miner will sometimes deliberately cross a stope under ground he knows to be bad, to avoid making a circuit when carrying his machine-drill to another place. He pays the forfeit.

The main levels are lighted by electricity, but tail-rope systems of mechanical haulage are used for traction because the distances are short and the tracks crooked. The horses employed for traction weigh from 1050 to 1200 pounds apiece and cost $200 at Seattle. They are lowered into the mine in a special harness, so made as to prevent them from kicking. When about to be taken into the mine, the horse is tied so that he cannot move and is then swung into the cage, with his head held up. Most of the horses are scared at first, but after a

ONE OF THE STAMP-MILLS AT TREADWELL.

couple of days they become broken to their work and learn to know what is wanted of them.

On returning to the surface from the gloom of the mine, the sunlight seems all the sweeter and the movement of life and color is keenly appreciated. The superintendent leads the visitor to one of the big mills, for instance, the one that contains 300 stamps and crushes 1650 tons of ore per day. The mills of the gods crush exceeding fine, but so does this one; and even the thunders of Jove would not silence the roar that comes from the batteries of the Alaska Treadwell. A stamp is like a hammer and it falls on a die that takes the place of the anvil. By dropping on the pieces of ore, the particles of gold are released, as a kernel in the nut that is cracked. The operation of crushing takes place within a closed iron box, called the 'mortar.' This has an opening guarded by a wire screen, so that the ore cannot issue until it has been pulverized. Water is fed into the mortar, and when the stamp falls this water is splashed against the screen. As it escapes from the mortar the water carrying the crushed ore runs down an inclined table covered with amalgamated copper plate. This arrests the gold, which readily combines with the mercury on the surface of the copper plate and forms an amalgam, that is, an alloy with mercury. While the gold is caught thus by the intervention of mercury, the pyrite, quartz, and other lighter minerals in the ore are washed to the foot of the tables and are led to vanners or concentrators on which further separation takes place, the worthless refuse flowing into Gastineau Channel while the heavy pyrite is saved, together with the fine gold closely associated with this iron mineral. The concentrate thus obtained is shipped in bulk by steamer to the smelter at Tacoma, where it is smelted with lead ore. The gold obtained as amalgam is retorted, that is, the mercury is distilled, leaving the gold behind in a spongy mass, which is finally purified by being melted with fluxes in a crucible.

Each stamp consists of a stem and a heavy shoe, the total falling weight being 1100 pounds. This mass falls 7 inches and 98 times per minute, so that it represents 62,883 foot-pounds or 1.9 horse-power. But much of the energy thus de-

ALASKA TREADWELL MINE IN WINTER.

veloped is wasted. Each stamp crushes an average of 5½ tons per 24 hours to a powder so small that it will pass through a wire-screen corresponding to a sieve having 400 holes per square inch. The ore contains $2.35 per ton and yields $2.15, the waste assaying 20 cents. This accumulates along the foreshore and forms a long white spit of sand. The cost of milling ranges from 12 to 22 cents per ton, the low figure appertaining to the newest mill. The total cost of mining and milling, together with general expenses, is only $1.35, so that a profit of 43% is earned. During the year ending May 31, 1908, the Alaska Treadwell mine produced 743,097 tons of ore, yielding $887,509 in gold from amalgam, and $736,636 from the concentrate, so that the total yield was $1,624,145. Of this not less than $577,493 was profit. This is a grand old mine. Since June 1890 it has yielded $22,359,934 in gold, of which $10,438,933 has been profit.

If opportunity offers, go into one of the mills at night. Then the thunderous power and insistent energy of the stamps are emphasized by the stillness of the sleeping earth. Two rows of electric lights illuminate the building brightly; the splash of the water, the movements of the mill-men, even the voice of a speaker a few feet distant, are apparently soundless, for the rhythmic reverberations of the mill drown them completely. And yet it seems a quiet place; the big noise kills all the irritating little noises of life, as small cares are drowned by a calamity.

CHAPTER V.

THE MEN IN THE MINES.

It is related that some fool who landed from a Seattle steamer was heard asking an acquaintance at Treadwell: "Where are the slaughter-pens and the swill boarding-houses?" Somewhere and somehow this ignoramus had got it into his head that the workmen employed in the mine were poorly fed, badly housed, and engaged in a dangerous occupation. The subject is interesting.

To keep account of the men employed, the following system has been adopted: Every man working underground is given a small brass tag, equal in size to a 25-cent piece, with a number upon it. This number is placed against the man's name in the time-book kept by the foreman; when the worker comes out of the mine he deposits his tag in a box; the shift-boss on duty takes the tags and hangs them on a series of hooks having corresponding numbers. These hooks with their numbers attached are arranged in rows upon a board hanging in the foreman's office. When the last cage-load of men reaches the surface every number on the board should be covered by a brass check. If one is missing, the shift-boss and hoist-crew are held at the shaft pending a search for the holder of the missing tag; he may have met with an accident or failed to come out of the mine. Twice a Slavonian was found asleep; on other occasions new hands have lost their way; but the most common source of trouble is the forgetfulness of men who walk off with their tags. On various occasions during the three months preceding my visit half a dozen workmen had been left in the Alaska Treadwell mine through losing their way or ignorance of the fact that it was 'quitting time.' These were all new hands. If a man walks off with his tag, a search is made for him at his quarters or at the boarding-house. It is probable that the men detained at the shaft by such stupidity do, or say, something picturesque. If a man is guilty of this blunder twice, he is 'fired.'

The management, in co-operation with the men, has organized a club. This is a commodious building on the shore and near the boarding-house. It includes a large billiard-room, a reading-room, a writing-room (paper and envelopes supplied), bowling-alleys, a dark room for photography, a barber's shop, ordinary bath-rooms and also a steam bath (towels and soap provided), a small circulating library, and an auditorium, capable of seating 500 and equipped with a stage and scenery. No charge is made for anything except barbering, but food and drink are not served, that being left entirely to the boarding-house.

The administration of this club is in the hands of a board of directors elected by the members; every employee must be a member, for each one is docked $1 per month on the payroll, from the manager to the nipper. The only ones exempt are the Japanese and the Indians employed in the mines. At present the membership totals 1500; these elect 21 directors, who choose a president, a secretary, and a treasurer. The last is usually the cashier of the Company, which originally provided everything at a cost of $28,000 and now furnishes water, heat, and light. The debt of the club has been reduced to $6000, equivalent to the amount for which it is insured. The profits are devoted to betterments and entertainments. In regard to these, the policy is to give the men what they want, such as an occasional prize-fight, minstrel shows, vaudevilles, amateur theatricals. Three dramatic companies have been organized by the employees and their women folk. There is a band of 25 pieces, the instruments being provided by the Club, and its reputation resounds throughout the Northwest. There is also a Fire Brigade consisting of volunteer firemen under a paid chief; the men are divided into six companies that meet twice a month at the Club to transact business and hold a discussion,

ALASKA PERSEVERANCE MILL IN SILVER BOW BASIN.

followed by an entertainment, for which the Company supplies the cigars.

The reading-room has a large array of newspapers and magazines. Any group of five men can ask for a paper; for instance, some Greeks complaining that journals of every nationality were on hand except their own, a Greek paper was ordered. In English there is everything from *The Spectator* and *The Nation* to *The Commoner* and the *Christian Socialist*. Magazines are apt to be taken from their covers; out of two dozen fully half a dozen are stolen each month. The Slavonians particularly are regular in removing the papers in their own language. Naturally, such trash as *Puck* and *Judge*, *Leslie's Weekly* and *Harper's Weekly* are popular. *The Nation* and *The Spectator* are supplied for the benefit of one or two of the more thoughtful. The first attempt to provide a library ended in the disappearance of the books. During six months, when the doors of the library were left open, 60 out of 250 volumes were purloined, including Drinker on 'Tunneling.' The only valuable books left were an ornamental Bible and a Webster dictionary. Now the books are locked up and can be obtained from the janitor by means of a signed card. At present the library contains 600 volumes, of which 100 are in use at any time. Among the popular authors are Ralph Connor, E. P. Roe, Mary J. Holmes, Wilkie Collins, and Dickens. But Kipling is "not at all popular," nor are Ruskin and Stevenson. It is a nice point whether popularity or disfavor be the greater honor.

A worker at Treadwell pays \$1 for the Club, \$1.50 for medical service, \$2 for his bunk, and \$25 for his board. This makes a total deduction of \$29.50 out of an average wage of \$100 per month. The company's boarding-houses are not expected to make money; in fact, they ran \$4000 behind last year. The men are not under compulsion either to sleep in the company's bunk-house or to get their meals at the company's boarding-house. Wages used to be \$2.50, with board and lodging at \$1 per day; now standard wages are \$3.50, with bunk and board at \$27 per month, as already stated. When a man is entered on the payroll, he is asked to state whether he wants to board down-town or with the Company. The cost in the

AN INDIAN CAMP IN SOUTHEASTERN ALASKA.

town is $1 per day for board, except in the cheapest Japanese restaurants, to which the miners rarely go.

In all these matters it is apparent that the management recognizes the fact that the workers in the mine are more than machines, requiring care and attention, apart from any humane sentiment. Not that such sentiment is lacking. Both the manager and his assistant are educated kind-hearted fellows, whose duty and pleasure it is to supervise the men under their charge with a proper mixture of sense and sentiment. Their policy is to encourage married men to come to Treadwell. Married men are more steady in their habits and less migratory. To attract the best class of men, the Company has built a number of attractive cottages for the use of such foremen and shift-bosses as are married, and also for other members of the staff possessed of a legal mate. These cottages now form a prominent feature of the picture presented by the settlement as seen on an incoming steamer. In their red paint, with a background of green scrub and forest, they give a touch of pleasant vividness to the scene. The houses are better built than the average suburban cottage and they are cheerfully perched on the slope overlooking Gastineau Channel. Some of them have six rooms, including a bath-room and an attic, the last being used for drying clothes, a convenience necessary in this damp climate. Electric light, steam heat, and water are provided, with complete sewerage. These houses cost $1750 and are rented for $12.50 per month. The smaller houses, costing $1000 to $1500, have four rooms, including kitchen and attic, but no bath-room. They rent for $10 per month. Each group of houses has a telephone connecting with the general system. The tenants are allowed some say as to the arrangement and details of the houses. Any man wishing to get a cottage applies at the office, and when a vacancy occurs he has the option of renting. If a new married foreman arrives, it is usual to build a house for him, rather than keep him waiting for his turn. Last year the Company spent $105,000 on cottages. Apart from the inadequate return on the expenditure involved, the Company gains by being able to retain its best employees. Viewed in a broad way, it is a good investment.

The miners work for 10 hours; mill-men, for 12. Those working underground come up for their midday or midnight meal, an hour being allowed for this purpose. They go to work at 7 and quit at 6. Nevertheless, a large proportion of the force works on an 8-hour shift, especially where continuous labor is necessary, as among those attending on the hoists, cages, crushers, trams, and chutes. In the mines it is impossible for one shift to relieve another immediately on account of the gas liberated from the large amount of explosives used in breaking the ore.

The force employed in the mines and mills is both heterogeneous and polyglot. An effort is made to adjust the ratio of races so as to have more than half English-speaking, dividing the remainder between Scandinavians and Slavonians, that is, between the peoples of northern and southern Europe. Those labeled Slavonian include Montenegrins, Rumanians, Albanians, Dalmatians, Herzegovenians, Croatians, in fact, all the immigrants from southeastern Europe, including the northern borders of Turkey. The least literate are the Montenegrins; they are big men but lazy and stupid. An effort is being made to get more Italians, especially Piedmontese, who come from the Val D'Aosta and the French-Italian border. They are splendid miners.

The Indians native to the district make good workmen. They belong to the Thlingit nation, inhabiting southeastern Alaska from Ketchikan to the Copper river. In the big open-cut or pit only Indians are employed, because they can keep steady while perched on narrow benches overlooking the cavernous hole. They work by day only. All the young Indians speak English well. They get instruction at the Silkoh mission and the schools for natives established by the American government 20 years ago. In the Treadwell mine from 60 to 80 of them are employed. They are mostly machine-men, that is, operating the air-drills; they work steadily all the year round, and receive the regular wages, $3.50 per shift.

All the employees in the boarding-house are Japanese, except the head steward and his assistant. The dining room is a clean and cheerful place, with long tables covered with white oilcloth. Imitation palms serve as a graceful decoration. The kitchen is a model of its kind. Dish-washing is done mechanically by a conveyor traveling in water that is heated by live steam to a temperature ensuring sterilization. A cold-storage room is provided. I saw the pies and bread prepared for the coming meal; they were as excellent as those obtainable in a good hotel. Flies are not a nuisance, owing to the coldness of the climate.

Cleanliness and despatch characterize the boarding-house system. Both are needed in providing for 1200 hungry men. By erecting the buildings over tide-water there is no trouble in getting rid of refuse; and in this the flocks of seagulls play a useful part. While the Treadwell boarding-house has been considered a model of its kind, the new establishment for the Mexican and Ready Bullion mines includes several improvements, mainly mechanical, such as a Garis-Cochrane dish-washing machine, a patent vegetable steamer, roll-warmers, plate-warmers, steam-jacketed stock kettles, vegetable-peelers, power-driven ice-cream freezers, and many other kitchen conveniences such as are found in modern hotels. Tea and coffee are made in copper urns by the wholesome method of percolation, so that the vile hard-boiled decoctions of the average mining settlement do not poison the good food. As an example of the variety of dishes served in these boarding houses, I quote, on the opposite page, the *menu* on the day of my inspection and the day previous.

Sherbet or ice cream is served once a week; green onions, lettuce, radishes, and oranges are provided at frequent intervals.

The bunk-houses are of two types. In one there is a corridor running the length of the building and into it two rows of rooms open, the general entrance being through a central door with a transverse passage. This form of construction is objectionable on account of the noise; the men loiter in the corridor; when going to and fro they disturb those who are sleeping. It is not practicable to restrict a house to men on the same shift, as there are frequent changes from one shift to another. The new bunk-houses are made so that the rooms are back to back,

SAMPLE BILL OF FARE AT BOARDING HOUSE.

June 26, 1908.

BREAKFAST.

Rolled Oats and Milk.
Beef Steak. Onions.
Corned Beef Hash.
Boiled Potatoes.
Eggs.
Hot Rolls. Flannel Cakes.
Bread and Butter.
Honey. Syrup. Coffee. Milk.

NOON LUNCH.

Oyster Soup.
Boiled Mutton. Pickle Sauce.
Frankfurter Sausage.
Cod Fish Balls.
Mashed Potatoes. Sauerkraut.
Green Peas.
Rhubarb Pie.
Tea.

DINNER.

Baked King Salmon
Roast Beef.
Baked Potatoes.
Sugar Corn. Boiled Beans.
Bread Pudding. Fresh Strawberries.
Cake and Tea.

June 27, 1908.

BREAKFAST.

Germea Mush. Milk.
Beef Steak. Pork Sausage.
Fried Potatoes.
Hot Cakes. Hot Corn Bread.
Bread. Butter.
Honey. Syrup.
Coffee and Milk.

NOON LUNCH.

Vegetable Soup.
Roast Beef. Brown Gravy.
Corned Beef and Cabbage.
Mashed Potatoes.
String Beans. Tomatoes.
Stewed Prunes and Rice.
Pie. Tea.

DINNER.

Boiled Beef. Horseradish.
Ox Tongue.
French Fried Potatoes. Lima Beans.
Lettuce Salad. Canned Peaches.
Cake and Tea.

each opening outdoors and with no hall-way of any kind, so that they are like elementary semi-detached residences.

This industrial community on Douglas island is a queer medley of races, creeds, and languages. The Slavonians and most of the common people of southeastern Europe when they arrive in America are polite to the point of servility, they doff their hats and scrape the floor with their heavy feet; soon they learn to keep their hats on their heads, often with a truculent slant, which is in strong contrast to the spiritless attitude assumed in their former habitat, where, in the presence of an employer, they have an air as if to say: "Excuse me for living." They easily get grouches and hatreds among themselves and their standard of fair play permits curious expression. A Slavonian will think nothing of getting behind a dump and throwing a 'rock' at another man. They rarely become naturalized citizens and they acquire the English language painfully. Out of 600 Slavonians only 4 or 5 are naturalized; they herd together, like the Chinese. When one of them has saved $500 to $1000, he returns to his native land and spends his money there; in four or five years he comes back and begins to accumulate once more. The Montenegrins claim to be warriors; they certainly are not workers, for they are accustomed to having the manual labor performed by their women. No Hindoos are employed at Treadwell; they offered, but were refused, because of possible complications with the other races. Certainly, it is a queer *olla podrida* of nationalities and yet in time, and in a comparatively short time, these diverse racial ingredients will be fused in the melting pot of American life and out of it will come a product as unlike the original material as the bullion that is obtained from the crude ore placed in the assayer's crucible. It will not be refined bullion and it is not ready for the best uses, but it is a product of definite value.

CHAPTER VI.

THE GLACIERS OF ALASKA.

Most of the tourists who travel in Alaska go to Ketchikan, Juneau, and Sitka; they are shown the Muir glacier; they call at Skagway, and, if the weather permit, they are taken in an excursion train to the summit of the White Pass, where they get a glimpse of the 'inside'—that vast hinterland whence comes the gold that has enticed civilization to transgress the Arctic Circle. In the southeastern portion of Alaska the tourist sees many glaciers; the indented coast is everywhere guarded by the protruding snout of a leviathan body of creeping ice; every river issues from the blue grottoes under the ice-fields, every avenue through the coast range appears to be filled by vast glacial stretches that block access to the other side. Thus the general idea of Alaska, as seen from the Portland Canal to Skagway, and from Haines Mission to Seward, is of a region invaded by glaciers, leaving a few picturesque islands and a narrow strip of shore on which Indians and white men gain a precarious livelihood by fishing and mining.

This impression is wrong. The glacier-infested portion of Alaska is only the southeastern coast; the far western shores of the Seward Peninsula are free from perpetual ice-fields, and once the traveler crosses the coast range, in going to Dawson from Skagway, or to Fairbanks from Valdez, he is in a region devoid of glaciers. Between latitude 56 and 61° north, for a distance of 500 miles and a width of 100 miles, the ice-fields prevail. North of 61° glaciers are less prominent as far as 63°, and still farther north they do not exist. The reason for this distribution is simple. The clouds rising from the Pacific are blown eastward against the coast range, and when they strike the snowclad summits their moisture is condensed and precipitated in the form of snow. This snow feeds the glaciers near the coast. On the other hand, within the interior of the country the altitude is low, ranging from 1000 to 1200 ft., there are no lofty mountain ranges, and the climate is particularly dry, so that the snow-fall is slight. Speaking broadly, the coast province of Alaska is mountainous, misty, and ice-bound, while the interior province is undulating, arid, and sunny.

Alaska covers an area of nearly 600,000 square miles, and of this total only 43,710 square miles appertains to the part known as Southeastern Alaska, the province usually assumed by tourists to be Alaska. The Cordillera or main mountain system of North America follows the Alaskan coast as far as Cook inlet and then forms the backbone of the Aleutian islands, sweeping westward in a broad crescent from the British Columbia boundary far out into the Pacific Ocean. On the northern side of this curved backbone is a region distinct in its climatic conditions. The mists of the coast do not penetrate the dry sunny atmosphere on the northern watershed; the short summer is intensely invigorating; the long winter is crisp and cold, but also marked by clear weather. The low mountain ranges rarely reach an altitude of 5000 ft., while the big Cordillera attains such heights as 17,500 and 20,464 ft. The slope of the vast hinterland is westward and it is drained by the great Yukon river, which flows through the very heart of it for a length of 2300 miles, emptying into Bering Sea. There the coast is low and marshy, with long beaches surmounted by the tundra. For eight months Bering Sea is ice-bound and the fog sweeps over the lowlands of the coast. This is also true, in lesser degree, of that part of Alaska nearest the States. While the scenery of the inland sea between Seattle and Juneau is lovely in the extreme, it is fair to say that appreciation is never dulled by seeing too much of it; clear days are infrequent; the traveler enjoys a day of rare loveliness and then is granted a couple of days of veiled modesty during which the mist hides the landscape so as to sharpen his desire for the uncovering when Nature is again in a complacent mood.

On June 24 Mr. Robert A. Kinzie took us to see the Taku

glacier. We left Treadwell on the morning of a day so misty as to be on the verge of rain. The tops of the hills were shrouded. Passing down Gastineau Channel we saw the string of cottages, offices, mills, and shaft-houses that mark the activities of the Alaska Treadwell, Alaska Mexican, and Alaska United mining companies, all of which exploit the same lode and are under the same technical direction. At the southern end of the settlement a large open-cut near the shore, and pointing under the Channel, suggested the fact that the workings of the Ready Bullion mine (belonging to the Alaska

TREADWELL, ALASKA.

United Co.) reach 1500 feet under the water. These workings are not allowed to come nearer than 300 feet from the surface of the rock, the intervening 'sea pillar' being left to protect the mine from flooding. The little railroad between the mines and the wharf can be seen edging the shore and passing over trestles until it ends at a group of cabins occupied by the Indian employees. They are charged a rent of $1 per month, but this they evade by quitting just before the month expires. Two or three families will pile into a single cabin and the Company is not particular about exacting its rental. As the launch glided southward we saw Sheep creek on the left, with an old sawmill and concentrator, to remind us of past mining schemes and of present litigation in New York. On the right, a tramway terminating at the shore indicated a mine on Nevada creek, but the forest had overgrown the tramway, as time had hidden the memory of an abandoned enterprise.

Passing the southerly extremity of Douglas island we turned into Taku inlet, skirting the mainland on the east so that we could see a small Indian village and its neighboring cemetery. Old graves are scattered in the brush close to the water; they are queer little structures like dog-kennels or doll-houses with picture-writing preserving the heraldic record of the incumbent. White and red paint gives this graveyard a chromatic gaiety, in contrast to the dark forest of spruce and the sombre canopy of mist. No totem-poles are visible. The natives do not bury their dead, but wrap them in their blankets, with their trinkets and weapons; thus equipped the body is laid in a box and a little house is built above it. The ground is rocky and digging is difficult, hence the custom.

This village is mentioned by Vancouver, who gave the name of Point Bishop to the adjacent cape. Rounding the point, we saw the first ice-berg moving down the Taku inlet, which is three miles wide. The mist had lifted, the light played on the floating ice, which shone white as sugar where vesicular, and a vivid blue where clear, both colors being doubled by reflection in the smooth waters of the estuary. It was now nearly noon and the fog had risen, uncovering snowy summits, grand mountains, and wooded slopes threaded by waterfalls. The air was still and the water smooth, we heard the call of the cataracts and the boom of the glacier; the clouds, the vivid green of the shore, and the dark woods were all reflected in the mirror over which we glided with tremulous speed. Many small bergs drifted past. The sunlight broke through the clouds and bathed the peaks and snowfields in matchless splendor. Soon we passed the snout of a dead glacier—the Windom. This is soiled by the dirt of the moraine and is but little crevassed. Turning the next point we faced the front of the great Taku glacier. We approached as close as possible, but no nearer than $1\frac{1}{2}$ miles, for the ice-bergs broken from the front of the glacier made navigation dangerous to our small craft. The engine was stopped. As the movement of the screw ceased, the silence intensified the beauty of the scene.

The front of the glacier spreads forward from a valley enclosed by high rocky slopes; looking into this valley we see that it issues from a vast amphitheatre in the high ranges, whence the river of ice can be traced to its source among the snow-fields half-hidden in the rising mists.

THE TAKU GLACIER.

What is a glacier?

A glacier is an ice river. The rate of movement depends upon the slope of its bed, the volume of ice, and the momentum resulting. Like a river, the movement is most rapid in the centre and at the top; this is due to the retarding effect of friction on the sides and bottom. Glacial ice is compressed snow and is formed wherever the snow-fall is so excessive as to cause compression to a viscous condition, permitting flow to a lower level. This condition is explained by the fact that water at the freezing point is transformed, under varying pressure, from the solid to the liquid state. Although brittle as glass and inelastic as granite, ice fuses at 32° F. under the atmospheric pressure of 15 pounds per square inch; if the pressure be increased, the fusion point is lowered, that is, the water will not freeze at 32° F., but will assume a liquid state. In a glacier the ice is subject to alternations of pressure causing transformation of the compacted snow from a solid to a liquid condition so that it acquires a viscous flow. Owing to the tension due to motion over an irregular rocky surface, the ice cracks, forming those fissures called 'crevasses.' As the glacier descends a valley the slopes on either side shed their débris upon the stream of ice. This débris may have been loosened by frost, rain, or snowslides; however formed, it rolls onto the edge of the glacier and creates a fringe, called a lateral 'moraine.' In describing glaciers the terms used are of French derivation because they originated in the western portion of Switzerland where French is spoken. When two ice streams meet, the inner lateral moraines unite and thus a medial moraine results. All the débris borne upon the advancing glacier is deposited at its front as the ice either melts or breaks away and the pile of rock thus formed is called a terminal moraine. While rock material is thus carried on the top of the ice stream, the glacier also moves gravel and boulders along its bed. Some of these fragments of rock are embedded in the ice as it advances and scratch the rock surface. Thus striations are made by the small pieces, and grooves by the large ones. These are parallel to each other, and indicate the line of motion. When glaciers recede or become extinct, by diminution of the snow-fall and change of climate, these marks on the worn surface of the rock will survive and testify to the agency that made them. They are the evidence of violent friction and powerful erosion; their formation is accompanied by attrition resulting in rock dust, which, mingling with the water running under the ice makes the muddy stream that issues at the front of every glacier.

A glacier is regarded as having two parts, known respectively as the 'accumulator' and the 'dissipator.' These two parts are separated by the 'snow-line,' above which the stream of ice is being constantly fed, while below the snow-line the stream moves ahead but never diminishes. In winter this dead line dividing the two stages of glacier existence is nearer the front and in summer it is nearer the head, but it always marks the critical stage of development. Let us apply these definitions: A fall of snow mantles the accumulator and begins to move down-stream until its lower edge crosses the dead line, where it melts and is dissipated. The upper edge of this snow-fall is buried by later snow-falls before it reaches

ANOTHER VIEW OF THE TAKU GLACIER.

the dead line and by that time the original snow is deep down in the body of the glacier. The rocks that fall upon the margin at the extreme head of the ice river also pass downward, to emerge only at the extreme foot, while those that drop upon the ice near the dead-line appear the more quickly and contribute to the lateral moraines. Indeed, this new conception, first suggested by Harry Fielding Reid, modifies our old views of moraines as well as our ideas of ice accumulation and motion.

During a geologic period immediately preceding the advent of man, portions of the earth were covered with vast sheets of ice, as the Arctic region is today. At one time it was supposed that one continuous ice-sheet reached southward from the North Pole, over northern Europe and northern America, but later scientific investigation proves that this was not the case. The ice-sheet did not cover the whole of the North, it was disconnected from the Polar cap, and existed in the form of great blankets, which, in North America, had three main centres of dispersion, namely, Labrador, Athabasca (in northwestern Canada), and the Cordillera, which includes the ranges now known as the Rocky Mountains and the Cascades. The ice moved not only southward but also northward from a general centre around Hudson's Bay. The fringes of these ice-sheets terminated in numerous glaciers that reached as far south as New York, Ohio, Illinois, Missouri, Dakota, Montana, and British Columbia. A broad belt across North America was thus buried, except where the mountain peaks held their heads above the blanket of ice, which was 2000 to 2500 feet thick on the level and even a mile thick in the deep valleys. When the climate became milder the ice-sheet retreated, that is, its front, in every direction, melted faster than it was fed, until only relatively small and local portions survived in the form of glaciers such as we see today. The formation of this vast body of ice was due in a general way to a period of excessive cold, but the distribution of it was determined by the prevalence of moisture in the atmosphere as modified on the one hand by ocean currents and on the other by the configuration of the land. Thus, the warm moist air of the North Pacific is arrested and cooled by the coast ranges so as to compel a precipitation of the snow from which an ice-sheet derives its origin.

Alaska was *not* buried under the ice blanket, the northern limits of which reached only as far as the seaward slope of the coast range. While, therefore, southeastern Alaska, the region now distinguished by glaciers, was under the Cordilleran ice-sheet, on the other hand, the interior, constituting the mainland portion of the country, escaped cold storage, save where Pleistocene glaciers existed in the high mountains. Then, as now, the interior was comparatively arid; the moisture blown from the ocean being arrested near the coast, and there was no

THE EAGLE RIVER GLACIER, NEAR JUNEAU.

Photograph by Winter & Pond, Juneau. Published by permission.

piling of snow adequate to form a persistent ice-sheet. Then, as now, southeastern Alaska was the land of snow and ice. Owing to the cold climate, the moist air moving with the Japanese current, and the high ranges fronting the coast, this part of America is still pervaded with ice, in the form of glaciers. Between Juneau and Skagway there are 20 glaciers in sight of those on board the passing steamer. In Taku inlet there are fully a dozen of them in a distance of 20 miles. Taking the entire sweep of the Alaskan coast from the British Columbia border to the first of the Aleutian islands there are fully 5000 glaciers, that is, there are a number of immense ice-fields debouching as numberless glaciers of varying size. This fact escaped early explorers, although Vancouver was not blind to it. Even the Muir glacier, in Glacier bay, which is now an object of interest to a host of tourists every summer, was not described until 1879, when John Muir and S. H. Young made an investigation followed by a scientific report. From a later survey made in 1886 by a party headed by G. F. Wright, it was ascertained that the Muir glacier is "a stream of ice 5000 feet wide and 1000 feet deep, entering the inlet at an average rate of 40 feet per day." At that time the front rose 250 to 300 feet above the water and the central portion of it was said to be advancing at the rate of 65 to 70 feet per day. This in itself is not incredible, because the Augpadlartok glacier in Greenland has been known to advance 100 feet in 24 hours, but there are good reasons for doubting the accuracy of Wright's surveys. In 1890 H. F. Reid found the ice-front 130 to 210 feet high, and the depth of water 720 feet, so that the total thickness of the ice was 900 feet. Excluding the wings of the glacier, it had then a breadth of 9200 feet, or ten times as much as its depth—a relation prompting Reid to observe that "rivers are generally much broader than this in comparison to their depth." This careful observer found a velocity of 7 feet per day; therefore, while it is probable that in 1886 the Muir glacier was moving more rapidly than in 1890, there is good reason for believing that Wright made a serious error, the real advance being at most from 8 to 10 feet per day. But it is melting faster than it is advancing. The loss of ice by melting of the surface is at the rate of 15 feet per annum. Between 1880 and 1890 measurements proved a recession of 250 yards per annum; the glacier being fed from a source that no longer supplies ice as heretofore.

The retreat of the glaciers in Alaska began 150 years ago; at that time the Muir had "a magnificent front, 6 or 8 miles across, though its height was probably not much over 300 feet. Large bergs must have broken off in great numbers and made Cross Sound difficult to navigate, which accords with Van-

ON TAKU INLET.

couver's report." This retreat is still in progress, the glaciers dwindling while the inlets enlarge. From Vancouver's notes, made 115 years ago, it is inferred that the front was fully 25 miles farther south, that is, near the entrance of the bay into which the excursion steamers now make their *voyages de curiosité*.

But all the glaciers of Alaska are not receding. The Brady has advanced since Vancouver saw it. In September 1899 the whole front of the Muir glacier, for five miles back, was broken by an earthquake. This earthquake started the Malaspina to life; the latter has advanced since then, crushing the trees that had grown up since its previous retreat. Recessions and processions of glaciers are due to local climatic changes and to geologic disturbances; as long as the supply of down-coming ice at the back is more than equal to the thaw and disruption at the front, the glacier gains; when the supply diminishes, the glacier shrinks.

The earthquake that shattered the Muir glacier was felt in Juneau. For eight years the floating masses of ice blocked Glacier bay for a distance of 14 to 18 miles, preventing any close approach. At the same time serious damage was done in Yakutat bay and even farther west. In Disenchantment bay the beach was raised 40 feet.

The face of the Muir glacier is now 160 to 180 feet high; this means about 1000 feet of ice below water-level, for the front of the glacier is floating, and when afloat there is nearly nine times as much ice under water as there is above. The density of ice is 0.92, while the mean density of sea-water is 1.02. Owing to the air imprisoned in glacier ice, the bergs lie with one-seventh of their mass above the water. The face of the Taku glacier at high water is 140 to 160 feet high, plus the difference of the tide, namely 24 feet, so that as much as 184 feet is exposed at low water. This indicates that the bottom of the ice-front is grounded. The manner in which the ice breaks also indicates this fact, for the blocks tip forward, permitting the more rapid movement near the surface of the glacier to gain on the slow advance on the bottom, so that as the ice cracks the big bergs tumble forward with a terrific smash. At the time of the high tides, as the flood advances, the air and water are forced under the ice; then the glacier splits, crevasses are formed, and bergs are detached.

In many cases the front of the glacier at water-level is hollowed so that the ice is undermined until a crack is formed. The water near its surface is from 38 to 40° F. and melts the ice. When a mass breaks off in front there is a sound as of a cannonade, and since these movements take place in localities undisturbed by human industry the noise is tremendous. It has been supposed that the explosive violence associated with the detachment of bergs from the glacier was due to the sudden release of air imprisoned in a vesicular condition in the ice; but this pretty theory is inconsistent with the facts of later scientific observation. The thunderous salute of the glacier is simply caused by the cracking of the ice. When freshly broken the surface is blue; after exposure it becomes white, because the ice is composed of interlocking crystals,

THE FACE OF THE GLACIER.

which refract and reflect the sunlight at the intersections. Owing to melting at the junctions of crystals their partings become visible, producing a heterogeneous surface, which breaks up the light, as snow does.

The bergs like stately argosies go seaward; they are not wasted. Some of them serve to cool the throat of the thirsty and perform other similar beneficent functions. The local fishermen catch them; in fact, they may be said to lassoo them. Such ice is worth a cent per pound. At Treadwell any berg that comes in sight is promptly arrested and put in storage for use in the compressor. Every lowering of temperature 5° F. means 1% saving in efficiency. As the compressor develops 800 horse-power, 3% means 24 horse-power, at $50 per annum per horse-power. The trade in ice was an important item more than fifty years ago. In 1851 a group of men at San Francisco contracted for 250 tons of ice to be shipped from Sitka at $75 per ton. In October of the following year the price was reduced to $35 and a new contract was made for 1000 tons annually for three years. Between 1852 and 1859, 13,960 tons were shipped from Sitka and 7403 tons from Kadiak.

Red patches on the glacier are due to the same cause as the red snow, which has astonished people in other regions. The discoloration is due to the presence of a minute and low form of vegetable life, the *protococcus nivalis*.

The Taku glacier is moving at about the same rate as the Muir; the front of it, nearly 200 feet high and a mile wide, is cleft by numberless cracks; these traverse the coiled mass of ice and give it a serrated outline. The steep shore on either side has been eroded by the grinding of the moving ice and the rock is bare, save for the clinging moss, streaked by the white filaments of cascades. On the eastern side the shore rises to a rounded mountain, 2400 feet high, separated by a low saddle from its counterpart. These two smooth hilltops probably constituted at one time a *nunatak* twin, upstanding through the ice when the glacier was bigger. On the western side the mountains are nearly twice as high—fully 4000 feet—and their summits are not glaciated, but jagged. On that side several small cirques or hollows indicate places from which tributary glaciers formerly descended; for even the immense body of moving ice now called the Taku glacier is but a shriveled remnant of a much larger mass that once pervaded this region. This ice-field from which the glacier issues is overlooked by peaks rising to 7600 feet. The Taku ice-river is 2 miles wide for a length of 8 miles, becoming constricted to one mile at the outlet; it is the protruding paw of an ice-field reaching for 30 miles, close to the British Columbian boundary.

IN
A
SNOW
DRIFT.

AN
ICE-BERG
IN
TAKU INLET.

Taku inlet affords the contrast between a dead and a live glacier. The Windom glacier is bespattered with dirt and the forest grows up to the edge; the ice is not moving. The Taku or Foster glacier is alive, it is moving, and as the ice enters the waters of the inlet it is broken with a roar that reverberates among the mountains.

When the throbbing engine of our launch was stopped, the silence was intense. The air was perfectly still and the water reflected the surrounding shore. We were close to the bergs coming from the glacier's front. White and blue, as the light played on them, they glided like stately sail-ships to the sea. Suddenly a huge splash, as of a leviathan bathing, indicated that a berg had lost its balance, by melting, and toppled over. Our launch rocked and the waves made all the other ice-bergs tremble. Other mysterious noises, both near and far, betokened restless movement all around us. When a small piece of ice breaks from a berg the equilibrium of the floating mass is upset; it splashes and threshes around like a porpoise. We heard the constant call of the waterfall. Then there came the boom of a cannonade and an echo like thunder. Surely something tremendous had happened. A crack had been formed, a mass of ice detached itself from the glacier, and a berg was created. And if these noises were vivid, the colors in the scene fairly shouted. The rock slopes are purple with moisture, the scrub and forest are massed in dark blue and gray, the ice-bergs are azure, the glacier is white, save where cracks show in bands of sapphire. And all these colors are repeated, with a thousand tints, in the water that is moving gently seaward.

The wind veered and we were in danger of being surrounded by ice-bergs; we had to proceed down the inlet, passing near enough to the Windom glacier to be able to see the crevassing along its broken back. Lifeless and decrepit it seemed in contrast to its brother glacier. The sunshine breaking through the mists flooded it momentarily with light, but vainly. Warm upon the dying glacier fell the gleam of living day. A dead glacier and an extinct volcano are types of power laid low, of youthful tempers disciplined by age and reduced to an equilibrium so perfect as to be incompatible with life.

CHAPTER VII.

THE SILENT CITY.

This is the story of a scientific fake. It was skilfully done, so that many were fooled for a long time. The perpetrator was Richard G. Willoughby, known to his friends as Dick and to the public as the Professor. He came to Alaska from South Carolina, where he had been a Methodist preacher. This was an avocation for which he was well fitted by the possession of a long white beard and a resonant voice. The Professor was a good talker and, among other accomplishments, he was a ventriloquist. When he left the South he went northwestward to the Cariboo and the Cassiar mining districts, and finally reached Juneau in 1881.

In 1885 Dick Willoughby brought news to the people of Juneau that he had discovered a wonderful mirage; it was to be seen above the Muir glacier. He described the vision as that of a modern city, with church-towers, large buildings, vessels in the docks, and people moving in the streets. The wonderful mirage had been seen by him on several occasions, but especially on June 21, the longest day of the year, when the sunlight was particularly strong. This story was repeated by him at intervals on his return from various prospecting expeditions, until 1889, when a sensation was caused by the statement that he had actually succeeded in getting a photograph of the "silent city." Great was the excitement at Juneau and throughout southeastern Alaska.

An association of local men was formed at Juneau for the purpose of exploiting the discovery and of selling the prints struck off Willoughby's wonderful negative. It was decided to investigate the phenomenon and to get more photographs of it. In June 1889 an expedition was organized. At the head of it were the Professor himself and a man named Minor W. Bruce. Bruce represented the *Omaha Bee* and other newspapers. He was an enterprising journalist of the irresponsible kind and made an excellent second to Willoughby. Bruce had come to Alaska to 'write up' the country and some of the business men of Juneau thought that he was well qualified to advertise both the Silent City and, incidentally, the mineral resources of the region. Even those residents of Juneau who were sceptical as to the mirage were alive to the fact that the story served as a good drawing card to attract the people from 'below,' that is, the dwellers in the States. Under these auspices an expedition was equipped to observe and photograph the mirage, which, so said the Professor, was due on or about the longest day of the year, known to astronomers as the summer solstice. The expedition set sail, proceeding down Gastineau Channel, around the southern end of Douglas island, up Chatham strait, and thence to the inlet leading to the Muir glacier. A few weeks later an excursion steamer, the *George W. Elder*, returning from a visit to the glacier, brought news that a member of the Willoughby expedition had come aboard in Glacier bay and had stated that on the day previous Bruce had gone forth over the glacier with his camera to take a shot at the Silent City, which, so Willoughby said, was about to appear. A fog had settled over the ice, and although Bruce's camera was found, he was missing. Not far away from the spot where his camera lay, there was a wide crevasse, and it was feared that Bruce while wandering in the fog had fallen into this crevasse. The young man who brought this news to the captain of the excursion steamer asked for ropes and grapnels wherewith to explore the crevasse. He also requested some provisions. These requests were met, with assurances of sympathy and interest on the part of the excursionists; and when the *George W. Elder* arrived at Juneau the news of the mishap created much excitement, not only in Alaska but also in the States, the fellow journalists of Bruce doing their duty nobly. This stimulated the demand for photographs of the Silent City; "they went like hot cakes."

Nearly a month later the expedition returned to Juneau and as it disembarked it was seen that Bruce had been found; his head was heavily bandaged and a boy was needed to lead him to his cabin. Evidently he had suffered. All the town was agog to hear the news. He was interviewed. His story was that when the fog enveloped him while crossing the glacier, he had tried to reach the camp, but wandered in the wrong direction, so that when the sun finally broke through the fog he found himself isolated from his party. While trying to find his way back, he became snow-blinded. To be blinded by

IN CHATHAM STRAIT.

the glare from sunlit snow is painful, as those who have suffered can testify. Bruce had to stop; he sat down on the ice under the shadow of a large hummock, where he was found next day. His companions had searched for him and had heard his call. This was a fine yarn. The expedition brought Bruce to Juneau in order that he might get medical attendance. Willoughby explained that it was then too late in the season to get a new photograph of the mirage. But the sale of prints from his first negative proceeded in a lively manner and the tourists came to Juneau to hear all about the wonderful phenomenon seen by the Professor.

As a matter of fact Bruce really was snow-blinded, but he soon recovered. About this time, in July 1889, another steamer, the *Ancon*, went to Glacier bay and many of the passengers saw the mirage of a single spruce tree above the Muir glacier. The 150 excursionists returned to testify to this fact and the news stimulated interest in the Willoughby legend. More prints of the Silent City were purchased. In the following winter Willoughby sold the original negative for $500 to a photographer at San Francisco.

A print from the original negative of the Silent City was given to me by a friend at Sitka, and is reproduced here, together with the portrait of the perpetrator of this colossal fake. The Professor is shown in the act of shooting at Nature in one of her wonderful moods. The Silent City looks like a large English town; the negative has been over-exposed and the outlines are dimmed. The trees in the foreground are leafless; evidently it is not midsummer, and yet the Professor claimed that he had obtained the photograph on June 21, for only on the longest day of the year was the mirage perfect. This little discrepancy escaped general notice. The negative was on glass, 8 by 10 inches; it had been poorly developed and *it did not fit Willoughby's plate-holder*, nor could it have been taken by his lens, which was a portrait lens. These facts were ascertained by my informant early in the game, and if he did not hasten to expose the fraud, it was because he liked the old Professor, he saw that the myth helped to bring tourists to Alaska, and he could not see what harm was being done to anyone, the credulity of the public being scarcely worthy of any particular protection. At Juneau people used to stand in a row waiting their turn to buy one of the photographs of the Silent City, and the demand occasionally exceeded the supply.

The truth is that in 1887 Willoughby happened to be at Victoria, on Vancouver island, and while strolling on the dock he saw a young tourist from Bristol, England, who was in the act of selling a photographic outfit, including a box of plates all of which had been exposed. The negatives, together with the outfit, were bought by Willoughby for $10. Among them was an over-exposed and badly developed picture of the city

THE SILENT CITY.
The photograph of the supposed mirage.

of Bristol. It probably reminded him of a mirage and of the optical effects seen above the glacier. His imaginative mind came to the aid of his loose morality and from the union of the two arose the idea of the photograph of a Silent City vibrating in the tenuous air of Glacier bay. During the excitement that followed the events in 1889 the American consul stationed at Bristol, while on a visit to San Francisco, happened to see one of the photographs of the Silent City on exhibition in a store-window and recognized it as Bristol. This fact was not generally known. Upon sending a print to my cousin, J. C. Hurle, at Bristol, he was kind enough to make enquiries concerning the date of the building operations at the cathedral, the towers of which are readily seen to be undergoing construction in the photograph of the Silent City, otherwise the City of Bristol. The Clerk of the Chapter testified that "the western towers of the cathedral were completed in 1888, when the capstone of the pinnacles was laid by Mrs. Norris." It was in 1887 that Willoughby got hold of the photograph, which evidently was taken before the work on the cathedral towers had been finished, probably in the winter of 1886. Willoughby used to say that as he saw the mirage in successive years the church-towers appeared taller, but he never explained why the trees were without leaves in June. On the back of the photograph of the Silent City is the following inscription, which is well calculated to stir the somnolent intelligence of a tourist:

The Glacial Wonder or 'The Silent City.'

"For the past fifteen years Prof. Richard Willoughby has been a character in Alaska as well known among the whites as he has been familiar to the natives. As one of the early settlers of old Fort Wrangel, in which his individuality was stamped among the sturdy miners who frequented the then important trading post of Alaska, he has grown with the Territory, and is today as much a part of its history as the totem poles are identified with the deeds of valor, or commemorative of the past triumphs of prominent members of the tribes, which their hideous and mysterious characters represent.

THE PROFESSOR AT WORK.
Willoughby and His Camera.

"To him belongs the honor of being the first American who discovered gold within Alaska's icy-bound peaks, but his greatest achievement, from a scientific standpoint, is his tearing from the glacier's chilly bosom the 'Mirages' of cities from distant climes.

"After four years of labor, amid dangers, privation and sufferings, he accomplished for the civilized world a feat in photography heretofore considered problematic.

"It was on the longest day in June 1888 that the camera took within its grasp the reproduction of a city, remote, if indeed, not altogether within the recesses of another world. The

'Silent City'

Is here presented for the consideration of the public as the wonder and pride of Alaska's bleak hills, and the ever changing glaciers may never again afford a like opportunity for the accomplishment of this sublime phenomena."

This queer rigmarole was the work of Bruce. Of course, Willoughby was not the first discoverer of gold in Alaska, although he was the perpetrator of a "sublime phenomena." Among his other discoveries was that of "coal-oil in chunks," namely, asphaltum. He was able to scare the Indians by his tricks as a ventriloquist and he passed among them in safety by utilizing this accomplishment. On one occasion he had a companion who wore false teeth and a glass eye; between the two of them they buffaloed the natives much in the manner of the Major in Rider Haggard's story of 'King Solomon's Mines.'

Willoughby died two or three years ago. He made a living by selling mining claims, clearing $1500 to $3000 each year by quick deals, for he had a plausible manner and was an entertaining talker, with a great fund of anecdote. Among the miners he was particularly popular, for they were impressed by his smattering of learning. Willoughby was for 25 years one of the living landmarks of Alaskan development and his memory should be preserved as a warning to the credulous.

It will be interesting to separate the grain of truth from

IN SITKA HARBOR.

Photograph by E. W. Merrill, Sitka. Published by Permission.

the chaff of charlatanism apparent in the story of the Silent City. What is a mirage?

A mirage is an optical effect by virtue of which distant objects are seen out of their real position. Light in traveling from an object to the eye of the observer passes through the air; this air is not always of uniform density; in a hot country the layer nearest the earth will be so heated as to be rarified; in a cold country the lowermost layer over the ground is condensed by contact with the ice or snow. Above this lowermost layer will come others in succession and these may be successively rarer or denser. Such layers of air serve as mediums for bending the rays of light out of their straight course, so that they proceed apparently from a new position. The result is to give a magnified or a distorted image or even to bring into view an object not otherwise visible. For example, the men on the whaling ships that cruise in the Arctic are reported to have seen Nome while still north of Bering Strait. Nome is a small town on the shore of Bering Sea and to the explorers in that remote corner of the world it is the outpost of civilization, a place for comforts not obtainable in the wilderness of ice and snow; in other words, Nome is as the sight of home. Sailors and fishermen that are steering for the roadstead off Nome will be astonished to see Nome pictures in the sky, real as life, while still so distant from it as to be normally out of sight. When this happens the air is still, the layer near the surface is chilled so as to be more dense than the average. Light normally travels in a straight line. If it passes from one layer to another of different density, it will be subject to deviation; it is possible for the variation in density in going upward to be of such magnitude that the light will follow the curvature of the earth, so that an object actually below the horizon will be clearly seen at a great distance, but in an elevated position corresponding to the direction in which the light is traveling when it enters the eye. If the distribution of density is such that the rays from the upper portion of the object cross those coming from the lower portion, the object will be inverted. Most of these effects can be observed by viewing objects through a bad pane of window-glass, that is, glass of

SITKA, WITH MT. EDGCUMBE IN THE BACKGROUND.

Photograph by E. W. Merrill, Sitka. Published by Permission.

unequal thickness, producing a result like that due to layers of air of unequal density.

In hot and arid regions, where sandy plains stretch forth to a low horizon, the lowermost layer of air becomes rarified by the hot ground, provided that no breeze stirs the atmosphere so as to mix the layers of unequal density. A condition of atmospheric calm is necessary for the formation of a mirage. Under such circumstances the prospector in Western Australia or Arizona will see a lake with trees reflected along its shore, and many a man half-crazed with thirst has seen limpid water where only an alkaline waste existed. Imagination comes to the aid of refraction and the brain persuades the eye that it sees things that do not exist. The mirage is due to an inverted image of the sky appearing beyond the portion of the plain visible to the observer. This inverted sky simulates a body of water, and if any object, such as a tree, happens to break the horizon, there is the appearance of a reflection in a lake. In cold regions the distribution of a layer of cold air high above the ground will cause the lower homogeneous layer of air to transmit an image in its true position, while the reflection from the upper layer yields another but inverted image of the same object. Many strange effects are produced and the strangeness of them is heightened by the imagination of the observer. A mirage can be photographed, but a hallucination will make no impression on a sensitized plate; a mirage is a true image of a real object; a hallucination is a condition of thought in a distempered brain; one is objective, the other is subjective.

What Willoughby really saw above the Muir glacier we can judge from what you or I can see there today. Mirages are not infrequent; the air above the mass of ice is rendered dense and the dense layer serves as a medium for the phenomenon of refraction. On sundry occasions he probably saw the hummocks and pinnacles of ice refracted and reflected by the overlying air until they seemed like the minarets and towers of a city not made with hands, or, by aid of his imagination, he even saw a resemblance to the church-towers and belfries of towns many thousand miles away from the Muir glacier. Un-

AN ALASKAN TROUT STREAM.

Photograph by E. W. Merrill, Sitka. Published by Permission.

loose the imagination of a man so fundamentally ignorant and so constitutionally visionary as the Professor, and something was bound to happen. The mirage looked somewhat like a city. When he bought the photographic equipment at Victoria and found a foggy picture of a city, *that* looked to him like the mirage. He looked at it again, and yet again, and the more he looked at the over-exposed plate the more the image upon it looked like his city of the mirage, until finally, by aid of a stimulant not unknown in Alaska, he came to the irrevocable conclusion that he had at last obtained the photograph of the silent city above the glacier. Having persuaded himself, it was easy to deceive others. The fake prospered amazingly. Two men knew the truth. One of them, whom we may call the Judge, measured Willoughby's plate-holder and satisfied himself that the photograph could not have been taken by the Professor. The other was Colonel Richard Dixon, a kindly old Southern gentleman who suspected a fraud; he went to the Judge on the quiet and asked him to "put him onto the game," so that he might enjoy the fun. The Judge trusted the Colonel* and told him what he believed to be the truth. Thereafter these two old jokers used to meet, compare notes, and enjoy the humor of the performance, which kept Juneau in the forefront of tourist interest and newspaper notoriety for many years.

---

*Col. Dixon was the recorder of the Harris mining district, having been elected in 1881. When asked how he obtained his military title, the Colonel answered: "My boy, I won that title at the Battle of Pork and Beans on the Fraser river in the early days." This brevet rank was won at least more worthily than that of the Kentuckian who derived his title from having married the widow of a Colonel!

CHAPTER VIII.

SITKA.

Sitka is picturesque and historic. This little trading post on Baranoff island stands on the shore of a waterway that is guarded by pretty islands; in front rises the lofty cone of Mt. Edgcumbe, its fires extinguished and its crater capped with the cold snow; in the background is Mt. Verstovia, the name recalling Russian rule. It means that the mountain is one *verst*, or 3500 feet, high. Sitka lies off the main line of coastal traffic and, being now no longer the capital, it has not much to give it importance in the way of business; but as a museum of Alaskan history, Sitka is unique. Being also a clean pleasant village set in exquisite scenery, it is a place every traveler should visit.

Sitka is the native name and means high land; Sheet-kah represents the Indian pronunciation. Incidentally, I may add that Yakutat was the first Russian penal settlement and derives its name from the Siberians or Yakuts who served as guards for the prisoners. The native name for Yakutat bay is Thlah-kah-cek, meaning 'the breeding place for hair seal.' The old settlement of St. Paul is now Kadiak; the steamship maps spell it Kodiak, which is wrong.

Sitka has a population of 400 whites, including 200 Russians and Russian creoles, that is, descendants from the first mixture of Russian and Aleut. In addition, there are 700 natives. The chief citizen is Sergius Kostrometinoff, called George, for short. He is a Russian by birth who was living in Sitka at the time of the transfer; hence his friends label him "an American by purchase." Mr. Kostrometinoff is extremely well versed in Alaskan history and to him I owe much of the information concerning the early days of the settlement.

Sitka was built close by the shore; the Russians were afraid to push inland on account of the Indians. The Russian houses were made of hewn logs, with low ceilings, weather-boarded outside, snug and warm. Double windows, and a big tile stove, more than sufficed to withstand the winter. The climate is milder now than it was a hundred years ago; there is more rain, less snow, and the glaciers are retreating.

The Russian settlement at Sitka was a fort and a trading post. Both were designed for business with the natives. The block-houses were built in Governor Etolin's time, between 1850 and 1854. The one that survives used to stand 50 yards north of the pond behind the Post exchange or canteen; an American officer, Major Campbell, took this block-house and placed it on its present site. The stockade that protected the trading post had a zig-zag course from the foreshore to the lake; and at each corner stood a block-house. The stumps of the posts of this stockade are still visible, although rotted and almost hidden by the grass.

The captain of the *City of Seattle* is so inconsiderate as to land us at 3 a.m. We are given the rare chance of seeing the awakening of Sitka. On every roof the ravens roost, like mourners for departed Russian glory. One or two of them lift an inquisitive glance and croak solemnly. The flapping of wings stirs the dreaming silence. Suddenly, at 6 o'clock, the bugler of the U. S. Marine Corps sounds the reveillé. Surely that inspiring clarion will awaken the town. It does not. At 6:10 the ravens utter raucous croaks and one of them flies away with a shrill scream. Sitka still sleeps. At 6:12 smoke curls lazily from a chimney on the main street. That looks hopeful. At 6:14 a childish treble is heard from an upper window. At 6:15 an alarm-clock goes off in the house opposite. At 6:20 two cats awake the incense-breathing morn with melancholy anthems, punctuated with expressions of vicious disapproval. At 6:25 the ravens drop from their perch and flutter restlessly. At 6:30 the bugler sounds another call and a suggestion of breakfast floats in the air. At 6:35 a workman strolls down the street with his dinner pail; simultaneously

THE ESPLANADE, SITKA.

Published by Permission

a distinguished citizen is seen going to bed. Sitka salutes the dawn!

After Sitka finally awoke, and breakfast had been obtained at the Hotel Baranoff, Turner and I strolled along the shore to the Indian River park. On the way we met a Russian priest with black cassock and long hair—an unpleasant anachronism in western America. Dominating the main street, the Greek church lifts its green cupola assertively. It is a hideous structure; not ever the kindly haze of romance can soften its ugly lines. It was built in 1848 and is deemed an antiquity. To men who speak of ten years ago as the 'early days' and look upon the time of the Californian pioneers as historic, 1848 seems long ago, for many things have happened since then. But a church sixty years old would in Europe be considered so new as to require apology.

We pass the museum (which we visit later) and several pretty cottages, some of them old log-cabins. The path is near the shore and the scene is full of charm. An Indian in his canoe paddles across the bay, a fisherman spreads his nets in the sun, ducks fly athwart the shimmering water, the sea swirls round the little islands, and the splash of the incoming tide echoes among the rocks.

Entering an avenue in the forest reserve, we reach the Park. It is a Government reservation covering a bit of virgin forest through which flows a trout-stream, the Indian river. Among the trees are totem-poles; these were bought from the Indians for display at the St. Louis Exposition (1904) and then returned, at the instance of Governor Brady, to adorn this park at Sitka. In a clearing four large totem-poles have been erected; these represent the corner posts of a chief's house; they are carved on the side facing inward and upon them the sill of the roof would ordinarily be placed.

According to custom, the totem stood close to the door of the chief's house; it bore his heraldic record. Although the uncouth carvings on the totem-poles suggest idolatrous worship, it is certain that the Haidas and Thlingits, who developed totemism, used it merely to represent family characteristics, and to symbolize qualities belonging to individual chiefs. It

TOTEM-POLES IN INDIAN PARK, SITKA.

Photograph by E. W. Merrill, Sitka. Published by Permission.

would be as reasonable to impute sinister ideas to the European who puts his crest on his ring, his linen, or his carriage. Like most savages, the natives recognized a supernatural power and ascribed human intelligence to birds and beasts; this led easily to the idea of transmigration and to mythical notions concerning the change of shape from men to birds or other animals, especially the lordlier species, such as the eagle, the bear, and the whale. Certain characteristics of their chiefs were symbolized by animals, which were carved on the totem-poles. Each family displayed the crest of its head-man and when they intermarried the totem recorded the fact, for example, that the eagle clan had mated with the wolf. Families having the same emblem are held to be blood relations, between whom marriage is forbidden. Each house shelters several families and the carving on the door-post signifies to what totem they belong. On the graves of the Haidas and Thlingits, other totems are placed, not so ornate as those that stand before the houses. The whale or a monstrous cetacean resembling the orca or grampus, is often portrayed on the grave-totems, to typify power and voracity. The bear, called *hoots* in Thlingit, is the crest of the Shakes family and was adopted to symbolize the bravery of their ancestors. On a Kake grave-totem the figure of a white man is carved to remind the Raven clan that their tribesman must be avenged. A raven is about to swallow a halibut, symbolizing the fate of the white man. This is not so bad as the latest development of the totem idea. In front of the house of an Indian chief recently deceased I saw the figure of a bear badly carved in white marble with gilded eyes, teeth, and claws. It cost $220. What is taste? an appreciation of what is fitting. But the Indian is no worse than the white man: at Dawson I saw a tombstone made of galvanized iron. There is no accounting for taste, or the want of it.

In the evening Sitka is not without diversions. There is canned music. Two gramophones—one in the Court House and one in the Marine barracks—enter into a contest, calling to each other over the diagonal of the parade-ground. As there is no traffic, the air is fully possessed by these mechanical songsters. The strains of 'Cheer up, Mary' answer to 'Walz me around again, Willy.' Mt. Verstovia and Mt. Edgcumbe, across the water, look on imperturbably. The smoke wreathes itself in blue whirls as it rises from the chimneys of the dreaming village; the mists are laid in long bands that belt the dark woodland; the water reflects the dying day. No footfall is

TOTEM-POLE AT SITKA.

heard. This is elysium. No, by thunder, this is only civilization. One gramophone calls to another: 'Keep a little cozy corner in your heart for me'; and 'Cheer up, Mary, there is a rainbow in the sky' peals forth from the opposite side. I feel sad. My mother always said I was not musical.

The church contains several remarkable pictures. Most of

these are oil paintings covered with a sheet of metallic silver, except the hands, feet, and heads of the sacred figures or *ikons*. One is an ivory carving, and underneath it is a painting on ivory overlaid with gilded plate. The halos around the sacred heads are made of gilded silver. Of the two pictures of St. Michael, the patron saint of this church, one is painted on wood and covered with gilded silver, while the other is on canvas. The most celebrated picture in the church is the Lady

INTERIOR OF RUSSIAN CHURCH, SITKA.

of Kazaan; this is kept under glass; as much as $25,000 has been offered for it. Everything in the church was carried in sailing vessels around the Horn, the Lady of Kazaan having been brought sixty years ago. Several of the pictures belonged to the first church, built in 1816 and destroyed by the Indians in 1852. In the cemetery of the Greek church, north of the present parade ground, there is a monument containing the cross of the church that stood on that site, at the time when the Indians rushed it and used it as a point of vantage in their

THE LADY OF KAZAAN.

attack on the Castle. The chapel of the Lady of Kazaan, in the present church of St. Michael, contains a picture that was saved from the old church; it hung out of reach of the Indians, who stabbed it in four or five places with their spears. The canvas shows the marks, although it has been repaired.

We were shown the marriage crowns; these are silvered, gilded, and decorated with Siberian stones. They are held over the heads of the bride and groom by the 'crown-holders,' who take the place of the groomsman and bridesmaid, in the procession around the altar. The wedding takes place in the middle of the church and lasts three-quarters of an hour. We were also shown the robes of the priests and of the bishop. The former are old and worn; the latter are handsome in gilt and silver, and were brought by the bishop himself when he came to Sitka from Russia five years ago. The big bible is bound in solid silver, gilded. Each corner of the book, and the centre, bears a painting on porcelain decorated with imitation brilliants. This bible is used only on special feast-days. The silver is stamped with the hall-mark 84. The eight pillars that support the dome of the church are solid hewn logs fully 20 inches in diameter. Large swinging candelabra and banners complete the decorations. The banners include an American flag, a Russian mercantile flag, and the standards of the religious societies connected with the church.

All the paraphernalia mentioned are interesting as relics of a passing era. They are less interesting to the European than to the American unused to the fripperies of medievalism. The totems of the Indian seem barbarous to the Russian priest; the veneer of silver and gold, with sham jewelry, and queer portraits of mythical personages, seem barbarous to the modern American; and the modern American's worship of the ticker and the tape will seem someday a queer form of idolatry to Macaulay's New Zealander.

In the museum the most interesting objects are the canoes, by aid of which the country was explored both by natives and by the Russians from the Siberian coast. The *kayak* of the Eskimo and the Aleut was called a *bidarka* or *baidarka* by the Russians. I saw one being built at Nome. It consisted of a frame of spruce ribs, over which is drawn a cover of walrus skin that has been sewn into one piece, the last stitching being completed when the skin has been stretched over the wooden frame. The skin is wet; as it dries it contracts and hardens so as to fit the frame closely. This boat is usually built to hold one occupant, who sits erect in a hole at the centre, the rim of which serves as an attachment for a waterproof covering or shirt, called by the Russians a *gamlinka*, which goes over the

A BAIDARKA AND ESKIMO.

head so that no water can enter the interior of the canoe. This covering is made of membrane obtained from the intestines of the walrus or seal; it is thin, light, and strong. When securely tied around the wrists and neck with cord made of walrus ligature, the boatman is well equipped to face the spray. The larger boat, able to hold a family, is called an *oomiak* or *uniak*. The Russians call it a *bidarra* or *baidarra*. It is flat-bottomed, and consists of a wooden frame tied with seal-skin thongs, over which the skins of seals are stretched after having been prepared, oiled, and sewn together with walrus thongs to hold them in place. At the mouth of the Yukon, while aground on the bar, we saw many of these aboriginal boats. Each boatman had several spears, used for sticking fish, and to these an inflated bladder was attached so that the spear would float in the water. Some of the boats were provided with large bladders, serving as corks to render the vessel unsinkable. The armor worn by the native warriors was also made from material of marine origin; it was a coat having three thicknesses of walrus hide, padded heavily at the shoulders. Speaking of armor, the cuirass of woven steel links worn for 27 years by Baranoff was found by Sergius Kostrometinoff in the possession of Shaketoo, an Indian chief. Mr. Kostrometinoff bought this coat of mail from the Indian chief and gave it to the Smithsonian Institution in 1906.

From a point of vantage in the Russian Lutheran cemetery, the traveler can obtain a good view of the Bay with its "thousand islands," the northern approach through Whitestone Narrows, and the wireless telegraph station on Japonsky island (where a Japanese junk was once wrecked). On the south shore are the Agricultural building standing on Castle hill, the Russian gray barracks now used as a Court House and jail; next to it, the Custom House and Post-Office; on the wharf, a big red warehouse; while near-by are the tops of the houses in the Indian village. To the east, half-hidden by the trees, is the old block-house and the new magnetic station; beyond are Silver bay and snowy peaks. Behind the town is Swan lake and Mt. Verstovia, with an intervening valley in which the experiment farm battles aggressively with the stubborn wilderness. Close at hand the graves with their Greek crosses are almost smothered by salmon-berry bushes and the rank vegetation of a brief summer, including the briar rose, now in bloom and bearing the perfume of other days.

CHAPTER IX.

HISTORICAL.

Let us turn back to the pages of history and seek the story of Sitka and the Russian occupation. In June 1741 Vitus Bering, a Danish captain in the Russian service, sailed from Kamchatka hoping to reach the American mainland. Ten years earlier some Cossacks, caught in a gale, had been driven across the Pacific to the shores of the eastern islands and had seen the continent of America. This had excited interest at St. Petersburg. On July 15, 1741, one of Bering's lieutenants, Alexis Chirikoff, anchored off the coast, near Cross Sound. This event marked the discovery of Alaska. Bering himself, an incompetent navigator and a court favorite, made no useful explorations, but hovered on the coast until his death in December, 1741. The furs his sailors brought back to Kamchatka aroused the greed of the Russians and led many of them to brave the crossing to the opposite coast. The fur-traders or *promishleniki* sailed the stormy sea in boats 30 feet long and 12 feet broad, with flat bottoms, made of plank fastened by walrus thongs and calked with moss. The sails were made of soft dressed reindeer skins, such as the Eskimo wear, and for ropes they had straps of elk skin. Thus a fur trade with the Aleuts was begun, and with it came the usual atrocities perpetrated by the semi-civilized adventurer when dealing with defenceless natives. Bancroft observes: "As the little sable had enticed the Cossack from the Black Sea and the Volga across the Ural mountains and the vast plains of Siberia to the shores of the Okhotsk Sea and the Pacific, so now the sea-otter lures the same venturesome race out among the islands, and ice, and fog-banks of ocean."

In 1779 the Empress Catherine II issued a *ukase* ordering the Aleuts, or inhabitants of the then known portion of Alaska, to pay tribute. In 1781 Ivan Golikoff and Gregory Shilikoff formed an association for the effective control of the fur trade. In 1783 Shilikoff erected a factory on Kadiak island, and in 1790 at Irkutsk he organized another fur company. In that year Alexander Baranoff, a sailor conspicuous for his energy, was put in charge of the trading post of Kadiak. He was soon appointed one of the directors of the Russian colonies.

In 1795 Shilikoff died. In 1799 the Emperor Paul gave the control of the Russian colonies to the members of the old fur company under the name of the Russian American Company, Alexander Baranoff was placed in charge and became the Chief Director or Governor. He ruled with a rough hand from July 27, 1791, to January 11, 1818. Astor's agent, Hunt, describes Baranoff as "a hyperborean veteran, overflowing with hospitality, who, if his guests do not drink raw rum and boiling punch as strong as sulphur, will insult them when he gets drunk, which will be shortly after he sits down to table."

On May 25, 1799, Baranoff established the first trading post at a spot six miles north of the present site of Sitka. This is now called Old Sitka, but the Russians called it Fort Archangel Gabriel. Nothing remains of this first settlement, but the natives use it as a fishing station in summer. The Russian church has placed a cross on the site.

Baranoff then returned to St. Paul, on Kadiak island. During his absence, in June 1802, the Indians massacred the Russian settlement, killing all the officers and 30 men. Only 5 Russians survived. It is claimed that the Hudson's Bay Company was interfering with Russian trade by selling muskets to the natives; the British traders would come as far as Lindenberg harbor (near Silkoh bay), sending an Indian in a canoe to tell the natives at Sitka, and the latter would then pack up their pelts and meet the Hudson's Bay factor, bartering furs for guns and ammunition. The Indians then lived on Crab Apple island, at the entrance of Whitestone Narrows. After the massacre, the natives moved to the mouth of what is now known as Indian river and built a stockade. From this stronghold they defied the Russians. In September 1804 Baranoff arrived with two ships and shortly after he was joined by Capt. Lisiansky, with the gunboat *Neva*. They anchored their vessels between Columbine island and the mouth of Indian river. The *Neva* opened fire with her guns and Baranoff made an unsuccessful assault

INDIAN RIVER PARK, SITKA.

upon the fort, being wounded himself. Five days later the natives evacuated the fort, because their ammunition was exhausted. They had killed their children and their dogs, lest by making a noise they might give the alarm when the retreat was made. Baranoff landed and found the fort empty, save for dead children and dogs, and one live old woman. He established a new post and called it the Fort Archangel Michael, and the settlement he called Novo Arkhangelsk or the New Archangel. This became the existing town of Sitka. Baranoff lived there. He built his first dwelling where the kitchen of the present Court House stands; later he erected a residence on the hill where the building of the Agricultural Department now looks out to sea. In 1813 he improved this house, which was finally reconstructed by Governor Kuprianoff in 1836. It was then called the Baranoff castle, and had two stories made of large logs, traversed by iron bolts between each window. On the top of this structure was a cupola, used as lighthouse, and in the basement a cellar for storing ammunition. In 1867 Alaska was transferred to the United States. On March 17, 1894, the castle was consumed by fire. It is suspected that the disaster was caused by an incendiary who wanted to destroy certain court records. A year before the fire the United States Government had repaired the building and made it suitable for the sittings of the District Court. The officials were just moving in, and the only occupant at the time was the U. S. Commissioner, Robert C. Rogers, an old man, who lived upstairs on the northwest side of the building. The fire broke out on the east side under the Judge's chamber, and on that very day the Judge had been examining certain accounts in which irregularities were suspected.

After the burning of the castle the present Marine Hospital was used as the residence of the governors of Alaska, until Juneau was made the capital. The site of Baranoff's castle is now occupied by the Alaskan office of the U. S. Department of Agriculture. This ornate colonial building was erected in 1902 and bespeaks the systematic effort to encourage the primal industry.

An army post was established at Sitka after the transfer, and persisted until 1877. The marines were first stationed here in 1880, and two companies are in residence today. They occupy the former hospital, the old Russian barracks being used as a jail. On the wharf is a large warehouse built of heavy logs; when the transfer from Russia was made, this building contained 30,000 seal-skins which were sold for $2.65 apiece.

THE BARANOFF CASTLE, BEFORE THE FIRE.
From an Old Photograph.

In the spring of this year (1908) seal fur sold for $25. The large log building on the right, going up the main street from the landing, was the warehouse of the Russian American Company.

The United States paid $7,000,000 for Alaska, and $200,000 for the property of the Russian company, including the castle, warehouse, barracks, and so forth. The smaller part of this transaction has always been involved in mystery. At the time of the transfer Prince Demetrius Maksutoff was Governor; he had occupied this position since 1864. Among those who landed from the steamer *John L. Stevens*, bringing the officials that took part in the ceremony by which the transfer of Alaska was effected on October 18, 1867, was a San Francisco merchant named H. M. Hutchinson. He proceeded at once to the castle, and made an arrangement with Maksutoff whereby he acquired the Russian company's vessels and other property for the firm of Hutchinson, Kohl & Co. Later, there was a sign on the door of an office in San Francisco reading the Maksutoff, Hutchinson, Kohl Company. When the Russian government learned of this, an objection was raised to the use of the Governor's name, and it was stricken out, but he retained an interest and is said to have made a lot of money by his participation. In 1869 the Alaska Commercial Company was incorporated with a capital of $2,000,000. In 1870 Congress granted a lease of the Prybiloff islands to this company for 20 years. In 1872 the Alaska Commercial Company purchased the property of Hutchinson, Kohl & Co. In 1876, a year before the withdrawal of the troops, a man named Whitford rented a portion of the warehouse and used it as a store. After the withdrawal of military authority he took possession of the entire building, until Brady, a missionary who was appointed Governor, joined Whitford under the name of the Sitka Trading Company. They were never dispossessed, and in 1888 a law was passed confirming their rights. At the time of the anarchy in 1877 the people of Sitka appealed for protection to the British authorities at Esquimalt, the U. S. Government having apparently left them in the lurch. It is said to this day that the Alaska Commercial Company was anxious to prevent interference from Federal authority, hence

RUSSIAN BLOCK-HOUSE AT SITKA.

the apathy of the department at Washington. The matter was never elucidated, and the disposal of the $200,000 remains unknown.

Alaska has had a variety of rule and misrule. After the transfer the district was under military control for ten years. Garrisons were stationed at Sitka, Wrangell, and Tongass. In June 1877 the soldiers were withdrawn and the region came under the administration of the Treasury Department. The Collector of Customs became the sole representative of Federal authority. When the troops were withdrawn the people of Sitka were left without protection. The Indians, thinking that the U. S. Government had abandoned the country, became troublesome. Within a week after the soldiers' departure, the Indians cut down the stockade and invaded every unoccupied Government house, removing the windows, doors, and partitions. A period of disquiet ensued. This culminated in riot in 1878. On February 6 the chief named Kaht-le-ahn had given a feast or *potlach*, accompanied by free drinking of *hoot-che-noo*,* and the Indians went on the rampage. The Kah-sat-tee clan, led by Kaht-le-ahn, murdered the American in charge of the Hot Springs, 18 miles from Sitka, on the same island.† These Indians were about to make an attack on the Sitka settlement when the Kah-quan-tan tribe, led by Ah-nah-hootz, came to the rescue of the white people. The Indians fought among themselves and this caused a postponement of the assault on Sitka. A steamer (the *California*) arrived next day and some of the settlers embarked, carrying the news to Victoria, British Columbia. Twenty days later (on March 1) the British warship

**Hootchenoo* is made from molasses, to which are added flour, dried apples or rice, yeast powder, and sometimes hops. A thin batter is made by adding water to this mixture, and when fermentation has taken place a sour, highly alcoholic liquor is obtained. It has an abominable taste and odor.

†The Hot Springs are south of Sitka. Even before the coming of the Rusisans, in 1805, the Indians used the thermal waters. Up to the time of the transfer the Russians maintained a hospital at the Springs, the magnesian waters having proved beneficial to sufferers from cutaneous and other disorders. The principal spring has a temperature of 154° F.

A SNOW-STORM AT SITKA.

Photograph by E. W. Merrill, Sitka. Published by Permission.

Osprey came to Sitka and protected the settlement until the U. S. gunboat *Alaska* arrived, on April 3. In June 1879 the *Alaska* was succeeded by the *Jamestown*, and thenceforth the territory remained under the protecting wing of the Navy Department until civil government was established by the Organic Act of May 17, 1884. Thus three departments of the Federal government, namely, the War, the Treasury, and the Navy, in turn administered this "non-contiguous possession" of the United States. These changes of administration were accompanied by lax government, prejudicial to business. In August 1877 there were only 15 Americans and 5 Russians at Sitka. Bancroft emphasizes the demoralization during the military occupation of the first ten years; the soldiers behaved disgracefully, and the settlement was full of disreputable people of both sexes. Sitka was "a grand house of ill fame." Even after the change to naval control, the affairs of this distant American colony were allowed to drift, and Alaska, as a whole, suffered for a long time from the predatory schemes of adventurers and politicians. Bancroft speaks of Alaska's "midnight suns in midsummer, her phantom auroras in midwinter, and her phantom government at all seasons of the year."

The settlements at Sitka, Wrangell, Juneau, and Skagway at different dates improvised various crude but effective forms of municipal government. At Juneau, mining regulations were devised on the basis of early Californian custom. The code of California had become the foundation of the mining laws of the United States. It expressed the conventions of a democracy pure and simple, for the regulations were passed in open meeting and the vote of the majority was final. The rules for locating claims and for holding possession were just, brief, and to the point. Public opinion made them effective and a sense of fair play made them respected. Equity was law.

The Organic Act created an executive and a judiciary, but omitted a legislature. The gap thus left was supposed to be filled by the declaration that "the laws of the State of Oregon, so far as the same are applicable and not in conflict with the laws of the United States and the Organic Act," should be the law of Alaska; but from the first a doubt has been expressed

INDIAN WOMEN SELLING SALMON-BERRIES, AT SITKA.

Photograph by E. W. Merrill, Sitka. Published by Permission.

by the courts as to application of these laws. This afforded a signal example of the difficulty of governing one community by reference to the laws of another. On October 1, 1906, the capital was moved from Sitka to Juneau, as elsewhere related.

For several years there has been agitation for a Territorial government with a legislature able to enact laws for Alaska. This would mean that gambling would be legalized and saloons would be operated under a low license, as is usual when a Territory is organized. On applying for Statehood these disreputable features are usually withdrawn. The Organic Act made gambling a crime, but it is not possible to get a jury to convict; the Marshal seizes the paraphernalia and thus stops "the music of the little rolling ball" and the shuffling of many cards. During the last two years the judges have enforced the law. No open gambling exists in Alaska. The same is true under the British flag in the Yukon Territory. It is believed that the grant of a legislature and Territorial government would give political control to the labor-union and saloon elements of the population. Under the government of a Territory the 30,000 people of Alaska* would have to assume the burden of maintaining law and order, and of protecting life and property over 580,000 square miles.

Crime, except counterfeiting and offences against the Customs, would have to be detected and punished by Territorial officials. At present the Federal Court has jurisdiction, both civil and criminal. The cost of maintaining the courts of the District of Alaska in 1906 was $587,000; in 1907, it was $490,000. Compare Arizona, a Territory, with Alaska, a Federal District: In Arizona in 1906 the United States Government, through appropriation by Congress, paid $89,000, representing only the salaries of the court officials and of the Court itself when sitting as a Federal department. Under the dual system all crimes against the laws of the Territory are tried on the Territorial side of the Court, and the expense thereof is borne by the different counties of the Territory. All offences against the laws

*In 1908 the population of Alaska consisted of 31,000 whites and 35,000 natives, besides a floating population of six to seven thousand miners and cannery men, who come to the country for the summer.

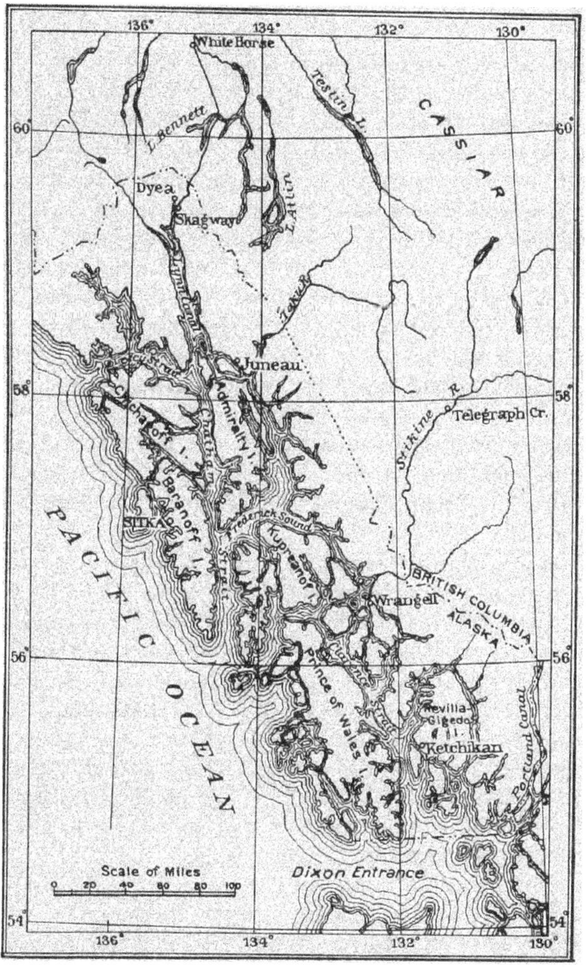

SOUTHEASTERN ALASKA.

of the United States are tried in the United States courts, this expense being debited to the Government. Offences against the United States include, besides counterfeiting and infraction of Customs, the selling of liquor to Indians, violations of the Edmunds Act, infraction of the postal laws, the revenue laws, and the like. The Government, of course, also pays the salaries of its own officials in the Territory and legislative expenses. The salaries of the judges are also paid by the Government, although the Territory usually sets aside additional sums for expenses of the judges when holding Territorial courts in counties where a Federal court is not held. In Alaska, the cost under Territorial government would have been $398,000 in the same year. Taxable property in Alaska is extremely limited; the land, except a few mining claims and townsites that are patented, is owned by the Government and pays no taxes. Agricultural land is negligible. The demand for Territorial government seems premature.

The most pressing need of the District of Alaska is a speedy and economical method of acquiring title to land. The system of United States surveys by which land is divided into whole, half, and quarter sections, has never been extended to Alaska, although for several years there has been an endeavor to engage someone able to make these surveys under the meagre appropriation allowed. In consequence, Congress has been obliged to enact so-called land laws for Alaska. These laws all necessarily contemplate a special survey by the applicant; therefore anyone at the present time desiring to acquire a title, whether mineral or agricultural, must employ a deputy surveyor, and have his survey approved by the Surveyor General, before he can buy the land. In all places where the system of Federal surveys has been extended, this preliminary work has been obviated.

Even the observations of a traveler warrant comment upon another matter pertaining to the administration of Alaska. Owing to the remoteness of the country and the scope of industrial activities, the scattered community needs judges, marshals, and other Federal appointees that have been carefully selected for the discharge of varied duties. It is foolish to try to govern the District from Washington, and it is criminal to appoint needy politicians to posts of unusual responsibility. A judge who has never even seen salt water before is sent to Alaska to decide important questions of admiralty jurisdiction. A man who is innocent of any knowledge of mining is appointed to adjudicate on fundamental questions affecting the development of the mining industry. Examples could be multiplied. Both as regards the fisheries and the mineral industry, the conditions obtaining in Alaska are unique; they demand a special system of rules and regulations, and they demand the services of men of approved character.

A few notes concerning the transfer of Alaska to the United States will be proper. Negotiations were commenced in 1861. In 1866 the Russian government refused to renew the charter of the fur company. Russia was unwilling to continue the expense of protecting a vast territory that was so unproductive, and she needed all her navy and resources to meet British aggression in Asia. Finally a treaty was arranged by William Seward, as Secretary of State, with the Russian envoy, Edward de Stoeckl, and it was signed at Washington on March 30, 1867. The United States agreed to pay $7,200,000 in gold.

On Friday, October 18, 1867, the steamer *John L. Stevens* arrived at Sitka; on board were the Russian commissioner, Captain Alexis Pestchouroff, and the American commissioner, General Lovell N. Rousseau, the latter being escorted by a company of the Ninth infantry. The 200 American soldiers, under General Jefferson C. Davis, marched up the hill and took a position on the east side of the flagstaff, which stood southeast of the castle. An equal number of Russian soldiers was drawn up west of the flagstaff. It was 3:30 in the afternoon. An account of the proceedings was given to me by Mr. Sergius Kostrometinoff, who, as a boy of 13, was present at the ceremony. All being assembled, Captain Pestchouroff ordered the Russian flag to be pulled down. The wind had twisted the flag round the ropes and by pulling them the flag was torn. This, as Mr. Kostrometinoff says, was a "pathetic sight" to the Russians.

A Russian soldier was told to climb the flagstaff and disentangle the flag. When half-way up he became exhausted and slid down. Another soldier failed likewise. Then a boatswain's chair was rigged up and a man was hoisted. His orders were to loosen the flag and bring it down with him; but he became excited and dropped the flag. The wind carried it away so that it fell on the bayonets of the Russian soldiers, at which they were visibly affected. In the meantime the shore battery and the American gunboat *Ossipee* were firing a salute. The Russian flag being down, the American was hoisted without any further incident. Captain Pestchouroff stepped forward and addressing General Rousseau, said: "By the authority which is vested in me by his Imperial Majesty the Russian Emperor Alexander the Second, I transfer the territory of Alaska to the United States." He spoke in English. General Rousseau replied, accepting the transfer. That ended the affair. "Thus, without further ceremony, without even banqueting or speech-making, this vast area of land, belonging by right to neither, was transferred from one European race to the offshoot of another." So says the American historian, Bancroft. It was an event of greater significance than anyone then living supposed, and it gave imperishable fame to the statesman responsible. Seward was severely criticized at the time, but he had been in the country and he believed that it had a future. "What, Mr. Seward," asked a friend, "do you consider the most important measure of your political career?" "The purchase of Alaska," he replied; "but it will take the people a generation to find it out." If the time of a generation be taken as 33 years, the fulfilment of his prediction was due in 1900. Assuredly Seward had been amply justified by that date, and he has been more than justified since.

CHAPTER X.

ALASKA AND CALIFORNIA.

The historic relation between Alaska and California is worthy of recital. It was founded on the fact that the one region lacked what the other produced, namely, grain and vegetables. If the Russians had been able to get supplies of green food in Alaska, their trading company might have done as well as the British East India and the Hudson's Bay companies. But the soil of the Alaskan islands and peninsulas was considered too sterile for cultivation, and the opposite shores of Kamchatka and Okhotsk were barren. Vegetables were scarce and scurvy was common. It was no wonder then that Baranoff welcomed the American vessel, commanded by Captain O'Cain, that brought a cargo of wheat and barley to the starving Russian settlement in 1803. This was the beginning of trade with California.

Before trade was established with the Spanish settlement at San Francisco, there was exchange between Alaska and the Sandwich islands, now known as Hawaii. Whaling vessels manned by Kanakas would put in at Sitka for fresh water and supplies. At Redoubt, 12 miles south of Sitka the Russians had a saltery and from it they used to send salt fish to the Sandwich islands, taking in exchange the brown or 'coffee' sugar. Several Kanaka words in the Chinook jargon serve as reminders of the early link between the cold shores of Alaska and the tropical islands of Hawaii.

After O'Cain discharged the cargo from his vessel, the Russian company under Baranoff had many dealings with American ship-masters and arranged for the sale of otter and beaver skins obtained outside of the company's possessions.

Baranoff supplied O'Cain with a party of Aleuts who caught otters along the coast from the strait of Juan de Fuca to the Golden Gate at San Francisco. The Spanish laws prohibited trade between Californians and foreigners, but the mission friars bought the otter skins under cover of night and paid for them in barley, wheat, peas, beans, and fruit. Baranoff's Aleuts took otter skins into the very Bay of San Francisco, the Spanish officials being unable to enforce their own regulations. And still the supply of food at Sitka continued scanty. The prospects of the Russian colony were gloomy, when, in 1805, Nikolai Rezanoff, the Imperial Commissioner, came to Sitka. Soon after his arrival he purchased the American ship *Juno* and her cargo of provisions. This prevented a famine. Then he sailed for San Francisco, with a view to making arrangements with the Spanish authorities for a regular supply of foodstuffs.

Unfortunately the Spanish laws were clearly against the exportation of grain. Don Luis de Arrillaga, the Governor of the Spanish colony, regretted that he had to enforce the ordinances of the King. He could neither sell grain nor buy the goods brought by Rezanoff on the *Juno*. Then came the romance of Concepcion, the 15-year old daughter of the *commandante*, Don José de Arguello. The Muscovite fell in love with the Spanish maiden; he was a manly and accomplished fellow; she was a beautiful and graceful girl. They were betrothed. Rezanoff told his sweetheart that he would die rather than go back to Sitka without food for his people, and the dark-eyed Concepcion assured her father that if her betrothed died, she would soon follow him to the grave. Thereupon Don José told Don Luis to regard him as an enemy if his obstinacy became the cause of a beloved daughter's death; and the friars all declared it was a flouting of Providence to deny them a market for their produce. The Governor perforce yielded. Rezanoff sold his goods and bought grain; the *Juno* was loaded with the necessary supplies and returned to Sitka in June 1806. Shortly after, Rezanoff returned to St. Petersburg, to obtain the imperial consent to his marriage. Being in a hurry, he went overland across Siberia from Kamchatka, and while

THE BAY OF SITKA.

Photograph by E. W. Merrill, Sitka. Published by Permission.

on his way was killed by a fall from his horse. His betrothed waited for him year after year, vainly, watching the Golden Gate for the ship that never came. Bret Harte tells us:

> "So each year the seasons shifted—wet and warm and drear and dry;
> Half a year of clouds and flowers—half a year of dust and sky:
>
> Yet she heard the varying message, voiceless to all ears beside:
> 'He will come,' the flowers whispered; 'Come no more' the dry hills sighed.
>
> Still she found him with the waters lifted by the morning breeze—
> Still she lost him with the folding of the great white-tented seas."

The world moved forward; on the ashes of the romance of Concepcion de Arguello, the missions in California established a regular trade with the fur-traders in Alaska.

Rezanoff left his mark on Russian policy. While dallying near San Francisco he wrote to the directors at St. Petersburg advising the establishment of a Russian settlement at the mouth of the Columbia river and another on the estuary of the Sacramento. "In this way," he said, "in the course of ten years we should be strong enough to make use of any favorable turn in European politics to include the coast of California in the Russian possessions. The Spaniards are very weak in these countries; if, in 1798, when war was declared with Spain, our company had had an adequate force on the ground, it would have been very easy to seize a piece of California stretching as far south as Santa Barbara." Apparently this suggestion was well received, for in October 1808 a vessel named the *Kadiak* was outfitted at Sitka for a filibustering expedition, under command of Alexander Kuskoff, a wooden-legged veteran. Loaded with rum, the *Kadiak* went to the Columbia river and did some successful trading. In 1810 Kuskoff went to California and was refused water by the Spanish officials. Cruising northward, he cast anchor in Bodega bay, 65 miles north of San Francisco. He reported a tolerable harbor, a fine building site, a mild climate, abundance of fish and fur-bearing animals, with friendly Indians and no Europeans. The Spaniards at San Francisco did not

SUN DOGS IN A WINTER SKY.

molest him, but they sent a protest to Madrid, and from there it was transmitted to St. Petersburg. The Russian Emperor Alexander I notified the directors of the Russian American Company that they might rely upon his protection. Soon afterward a convention was signed between John Jacob Astor and the Russian company, whereby an exchange of furs for provisions was arranged.

Surveys and negotiations concerning the settlement in California consumed a year. In 1811 a tract 18 miles north of Bodega bay was bought from the Indians for 3 blankets, 3 pairs of breeches, 2 axes, 3 hoes, and some beads. There was no anchorage, but in other respects this locality was preferable to the one first selected. Here in April 1812, just as Napoleon was preparing to invade Russia, a party of 95 Russians, of whom 25 were mechanics, landed. They were accompanied by 80 Aleuts. All hands set to work forthwith to fell trees for building purposes. By the end of September Kuskoff had erected a fortified village on a bluff 110 feet above tide-water and about eight miles from the mouth of the Slavianka, called San Sebastian by the Spaniards, and now known as the Russian river. The stockade and block-houses resembled those built by Baranoff at Sitka. A rectangular enclosure, 250 by 300 feet, was formed with posts 12 to 15 feet high, surmounted by a bar in which were set obstructing spikes of wood and iron. Hexagonal block-houses guarded each corner and in them cannon were mounted. This fortified enclosure was strong enough to deter Spanish attack. Outside of the stockade the Aleuts had their huts, and close to them were the windmill, granaries, cattle-yards, tannery, and workshops. A well kept garden adjoined these buildings. Thus the settlement was strongly protected and intelligently planned. It was called Fort Ross, or Russian Fort. *Ross* is the root of *Rossia*, the vernacular for our word 'Russia.'

The Spaniards disliked this Russian trespass into their sphere of influence and annoyed the new settlement as much as possible, but they could not stop the contraband trade maintained between Fort Ross and San Francisco. The authorities were compelled to wink at the infraction of regu-

Photograph by E. W. Merrill, Sitka. SITKA AND MT. VERSTOVIA. Published by Permission.

lations. In 1817 the *padres* founded the mission of San Rafael; this seemed to menace the Russians, but they sent gifts to the church. Soon afterward Canonigo Fernandez came from San Francisco as a representative of Mexico and notified the Russian commander that he must evacuate Fort Ross within six months. The commander glanced at his fortifications and told the envoy that the region had not been in possession of any other power when the Russians occupied it, that the site had been bought from the Indians, and that he was quite prepared to meet force with force. A similar attitude was assumed by Kostrometinoff, who was the next commander.* Nothing happened; possession was retained, trade was maintained, and the Russians began to have a title by prescription.

In 1831 Baron F. P. Von Wrangell was appointed Director of the Russian American colonies. He was an Arctic explorer, a scientific authority, a statesman, and a sailor; in fact, Wrangell was much the highest type of man ever connected with the Russian settlements in America. He visited Fort Ross in 1832 and it devolved upon him to decide what to do with this isolated possession. Being a diplomat, he began to fence with the Spaniards. To force the position, he established a trading post at Sausalito and negotiated for the cession of San Rafael to the Russian company. The Spanish governor retaliated by planting a settlement at Sonoma. Nevertheless, Wrangell was not enthusiastic over the future of Fort Ross. He reported to his company that unless they annexed the country eastward as far as the upper Sacramento valley and southward as far as San Francisco bay, they had better abandon Fort Ross entirely. It was costing 45,000 to 70,000 rubles annually, while the revenue from furs and other products ranged between 8000 and 25,000 rubles only. Another authority states that between 1825 and 1830 the expenses of the Ross settlement were 45,000 rubles, while receipts averaged less than 13,000 per year. The Russian company tried to persuade the Government at St. Petersburg that it would be well to secure a slice of California before it changed hands, but Nesselrode

*And uncle of the Sergius Kostrometinoff, of Sitka, to whom reference has been made.

was unwilling to embroil his country with the United States and turned a deaf ear to the proposal. [California was ceded by Mexico to the United States by the treaty of Guadalupe Hidalgo, signed on February 2, 1848.] Thereupon Kuprianoff, who succeeded Wrangell in 1836, suggested the abandonment of Fort Ross. In April 1839 the directors passed a resolution to this effect. Kuprianoff offered the Russian property in California to the Hudson's Bay Company for $30,000, and then, when they declined to buy, he offered it, in 1840, to the Mexican government, which also refused to consider the purchase. Thereupon Captain John A. Sutter appeared on the scene. He was in the habit of buying property on credit and he made a proposal to the Russian officials. On December 13, 1841, a formal contract was signed in the office of the sub-prefect at San Francisco between Sutter and Kostrometinoff, by which the latter assigned to the former all the property at Fort Ross and Bodega. This included 1700 head of cattle, 940 horses, 900 sheep, besides improvements and implements. The price was $30,000, divided in four annual installments: two of $5000 each, payable in wheat; a third of $10,000, also in wheat; and a fourth of $10,000, to be paid in cash. The sale included all the improvements, but not the land. On the day before the deed was signed, the manager, Rochef, executed a private deed assigning to Sutter for $30,000, the receipt of which was acknowledged, all the lands held by the Russians. This part of the transaction was kept secret, but nearly twenty years afterward, when property on Russian river became valuable, the deed came to light and many ranch-owners paid Sutter for quit-claims. The Russian government never asserted title to the land at Fort Ross, nor was there any reference to the subject in the negotiations preceding the transfer of Alaska to the United States.

The Russian garrison at Fort Ross embarked on the ship *Constantine* in February 1842. A single Russian remained as watchman until the arrival of John Bidwell, who assumed charge in behalf of Sutter.

The Russian American Company found it difficult to collect its price from Sutter. He was an impecunious person.

For three years no payment was made, either in wheat or money. Then a small contribution was made in the form of a consignment of wheat. It was believed at Monterey, which was the Mexican headquarters, that Sutter was negotiating for the transfer of his estate, then called New Helvetia, to other Americans; to prevent this, it was proposed by the Mexican authorities to pay off the Russian claim and acquire the mortgage on Sutter's property. These plans were never consummated, for soon afterward California passed under the American flag. The Russian company recorded its mortgage, but the influx of population following upon the annexation and the discovery of gold, in 1848, set Sutter on his feet. Before 1850 he had liquidated his debt. The last installment of $15,000 was paid to the company's agent at San Francisco, but the man absconded. In consequence of this theft and the expense of prosecution, a deficit appeared on the Russian company's books for 37,484 *rubles* and 50 *kopeks*. Thus ended the Russian occupation in California.

Sutter's name was made famous later by becoming connected with the discovery of gold in California. He was of Swiss parentage, born in the Grand Duchy of Baden, whence he emigrated to New York in 1834, at the age of 31 years. Arriving in California from the Sandwich islands, now called Hawaii, with a company of Kanakas, he became naturalized as a Mexican and obtained a license to settle in the valley of the Sacramento. At the confluence of the American and Sacramento rivers he established a colony called New Helvetia, on the site of the city of Sacramento. There he built a stockade or fort. As a naturalized citizen of Mexico, in 1842 he was called upon by the Spanish governor to oppose the invasion of California by the bands of Americans then beginning to come overland to the Pacific slope from Missouri and Arkansas. But Sutter was not solicitous for the integrity of the Mexican territory. He realized that it was to his interest to welcome, rather than oppose, the newcomers; he saw that they were destined to become the masters of California, and that their friendship was more advantageous to him than their enmity. Thus he excited the suspicion of the Mexican officials, who, however, were helpless to discipline him at this juncture. In 1844 he organized a military company, receiving the appointment of captain at the hands of Micheltorena, the Governor of Alta California. These preparations were against Alvarado and Castro, who had started a revolution, at a time when war with the United States was imminent. Micheltorena and Sutter united in fighting Alvarado and Castro. The former were

IN QUIET WATERS.

accompanied by a number of Indians and took with them the Russian cannon from Fort Ross. The opposing forces met at Cahuenga, near San Fernando, close to the present city of Los Angeles, in February 1845. Micheltorena and Sutter were easily beaten, and capitulated. Sutter was permitted to return to Sacramento, a sadder and a wiser man. But fate was kind to him; he was destined to win a distinction through which this fiasco would be forgotten. When war broke out between Mexico and the United States in 1845, the American flag was raised at Monterey on July 6 and at Sutter's fort on July 8, and from that date it has been the sign of American dominion in California.

Sutter served as Indian agent under the American administration, and in 1847 he was appointed special commissioner by Governor Mason. In the same year he did himself credit by promptitude and generosity in fitting out an expedition to relieve the party of 80 immigrants from Illinois, of whom 36 perished in the Sierra Nevada near Donner lake. Sutter behaved well; but his great opportunity was coming. His fortified enclosure on the Sacramento river had become a large establishment, over which he presided in a patriarchal way. Before the country was wrested from Mexico, this enterprising Swiss had power to inflict punishment and he was monarch of all he surveyed; when American settlers poured across the mountains from the East he saw that there would be a good market for lumber, so he planned the building of a saw-mill on one of the streams issuing through the foot-hills of the Sierra Nevada. Among others sent by Sutter to search for a suitable site was James W. Marshall, a carpenter from New Jersey. This was in the summer of 1847. Marshall returned in a month saying that he had found a suitable spot on the south fork of the American river, at a place now known as Coloma, about 35 miles northeast of Sacramento. It was arranged for Marshall to build and run the mill, which was erected by the middle of January 1848. When ready it was found that the ditch or race, leading the water from the wheel, was not deep enough. Marshall scoured it with the swift current, opening the flood-gates to full capacity. The water was allowed to run all night, and in the morning the gates were closed while Marshall examined the mill-race. In the gravel loosened by the current he saw several bits of gold. By hammering a specimen with a stone he ascertained that the heavy yellow metal was gold. That was on January 19, 1848. A few days later he went down to New Helvetia or Sutter's fort and told Sutter that he had discovered gold. Sutter tested the metal with aqua-fortis or nitric acid, which he found among his apothecary stores; he read the article on gold in his copy

SITKA AND MT. EDGCUMBE.

Photograph by Winter & Pond, Juneau. Published by Permission.

of the Encyclopedia Americana; he weighed the metal and compared it with coins; thereupon he pronounced the substance to be gold and nothing else.

That incident marks the beginning of the Golden Age of California's youth. A wide-spread mining excitement followed and fortunes were made by thousands of men. But neither Sutter nor Marshall had any legal claim to the ground on which the gold was found, although a few months later Sutter sold his supposed rights for $6000 and Marshall disposed of a one-third interest in the mill and timber for $2000. Neither Sutter nor Marshall benefited by the historic discovery with which their names are indelibly associated. In 1870 the Legislature of California passed a bill appropriating to "General John A. Sutter, the pioneer of 1839 and founder of New Helvetia, once the richest and most powerful foreigner in the country, but by that time reduced to poverty, a sum of two hundred and fifty dollars per month." I quote Hittell. A similar grant for Marshall was passed, but not being approved by the Governor, it failed to help the poor man who had opened the natural treasury of California. In 1872, however, an appropriation of $200 per month for his support was granted and was paid to him for two years; and in 1874 he was voted $100 per month for the next two years; that was all he ever got from the State. In 1885, at the age of 73, he died alone and in poverty. After his death, in 1887, the State, at an expense of $5000, erected a monument to his memory on the spot where he first found the gold. Such is the irony of fame.

CHAPTER XI.

CHINOOK, NATIVES, AND GAME.

In New Zealand and in Hawaii the natives have been decimated in a century; in Tasmania the aborigines are extinct. The disappearance of native races is due in part to the introduction by the white man of new diseases that attack the natives with great virulence, and it is traceable in part to alcoholism induced by the excessive use of stimulants sold to the natives by traders. The chief factor, however, in destroying the vitality of the Maoris, the Kanakas, and the Haidas has been the imperfect adaptation to a new environment created by the introduction of civilization. Thus the Alaskan Indian, becoming employed by the white man and earning wages, adopted the white man's food without donning his warm clothing. Previously, the healthy savage had fed on fish and seal-oil, especially oil, which, on account of its heating quality, enabled him to withstand extreme cold and excessive damp. When he came within the white man's camp he bought cakes, biscuits, hard tack or pilot bread, and similar non-heating food. That made him less robust and predisposed him to tuberculosis. He continued to wear a shirt and drawers made of cotton cloth, over which he threw his blanket. When hunting he wore moccasins; otherwise his feet were bare. Nowadays, however, many of the Indians, especially around Sitka, wear the clothing of the ordinary white laborer and they are the healthier for it. By living indoors they have become soft; to protect themselves against the effect of a change in food, they must be better clothed. Those that do so, are increasing; those that accept part of the gifts of civilization without its burdens, are dying of lung trouble. Westward and in the northern interior, among the Eskimo, the conditions of life are still as they were in southeastern Alaska twenty years ago. When the native tribes first come in contact with the advancing wave of modernism, they wilt. When they change their clothing and their dwellings in complete acceptance of new habits and conditions of life, they survive. It is an interesting example of natural selection as influenced by adaptation to environment.

The medicine men of the Thlingits are called shamans, and like the priestly caste of other peoples they made trouble. When chant or hocus-pocus failed to cure the sick, they demanded the death of another whom they charged with bewitching the invalid. Thus they vented their spite on enemies. It is related that Captain E. C. Merriman of the American navy destroyed shamanism in southeastern Alaska by capturing some of the shamans, taking them on board his ship, cutting off their long sacred hair, and sending them bald to their tribe, by whom they were received with uproarious laughter. Ridicule killed their black art.

In their dealings with white men, the Indians speak Chinook. This is a jargon composed of many tongues. When the *voyageurs* and *coureurs de bois* of French Canada traveled over the wilds of the Northwest, they made acquaintance with many tribes of Indians speaking different languages; they also traded with the Scotch factors of the Hudson's Bay Company. Thus they gradually gathered a composite speech containing words of French and English origin, as well as sundry words picked up from native tribes, such as the Kictatats, Haidas, and Thlingits. After a certain number of mixed words had come into use, the Hudson's Bay factors went to work and developed this lingo, called 'Chinook,' so that it might serve them in dealings with the various tribes from whom they bought furs and fish. Chinook consists of about 300 words and is easily learned; it has a vocabulary, but no grammar. Most of the words are corruptions. Thus 'siwash' is the general term for Indian throughout the Northwest; it comes from *sauvage*, just as 'musher' is corrupted from *marcheur*. 'Mush on' is probably a corruption of the French *marchons*. *Klahowyah*, the native salutation, is said to be derived from "Clark, how are you?" the greeting given to the old-time traders. 'Sour dough' and 'cheechako' are complementary. 'Sour dough' is the emblem of the seasoned frontiersman. Being unable to procure yeast, the prospector or woodsman carries a little can filled with soured dough-batter; with this and by the addition of a little baking soda, he starts the leavening of his bread, in the form of pancakes, or 'flapjacks.' The men of the North will allow the lump of sour dough to freeze and as the stock is diminished they add

THE MUSHER.

flour and water, mixing the mass, so that it performs for them the function of yeast. 'Cheechako' or 'chichaco' is probably of Kanaka origin. *Chi* or *chee* means 'new' and *chaco* or *chako* means 'to come'; a 'cheechako' is a newcomer. The term corresponds to 'tenderfoot' in the West and 'new chum' in Australia.

Chinook is composed of derelicts from English, French, Indian, and even Kanaka. Long before the Russian or the

American controlled Alaska, there was trade between the Russian colonists and the islanders in the South Seas. But that is another story. *Klootchman* is the Chinook for native 'woman'; in the Thlingit language it is *sha-wat.* *Skookum* is the word for 'strength' of any kind; thus *skookum tumtum* is 'strong heart' or 'courage'; *skookum sick* means 'very ill' or 'about to die'; *skookum koolie* is 'to go fast.' *Koolie* is probably from the French *coulé,* which also appears in our word 'gully.' *Cultus* is Chinook for 'worthless.' If you say a man is 'cultus' you mean he is worthless, a 'waster,' or *vaurien.* *Cultus wawa* means worthless conversation. If you joke with a native or tell him a story he cannot believe, he will laugh and say: "*Cultus wawa.*" As in most primitive languages, emphasis is obtained by doubling; thus the Australian Yarra Yarra is the aboriginal equivalent of 'ever flowing.'

As an example of the development of local jargon, I quote the expression: "to siwash a line." To explain it, I shall describe an interesting bit of pioneer engineering. The Alaska Treadwell company is building a dam in the natural basin or cirque two miles south of the mine and under the shadow of Mt. Jumbo; the dam is to be 60 feet high and will impound 240,000,000 gallons or 34,000,000 cubic feet of water, sufficient to afford a flow of 200 miner's inches for 78 days. The area to be transformed into a reservoir covers 35 acres. In undertaking the construction of the dam, the first thing was to get a donkey-engine to pull logs, for the mossy nature of the ground precluded the use of horses. The machine consisted of a horizontal engine having a 9 by 10 inch cylinder with a vertical boiler, both boiler and engine being set on a solid cast-iron base, placed upon a heavy sled. The sled was built of two 16 by 28 inch timbers, faced on three sides, the bottom having its original log surface. Each end of these 'runners' is 'sniped off,' so as to present a slanting point like the prow of a boat.

At first sight it seems absurb to think of pulling such a machine, weighing 10 tons, over the moss, morass, and rock of a primeval sub-arctic forest. But it was done, thus: A light cable is paid out to some suitable mooring, such as a tree, and

THLINGIT WOMEN.

Photograph by Winter & Pond, Juneau. Published by Permission.

then a block is fastened to the same hold; the line is placed in the block and the end of the light line is attached to the block of the heavy line. The donkey-engine pulls in the light line, thereby pulling out the heavy cable; the block of the latter is attached to a tree, either the one already in service or a larger one, or the cable is even run around two or three trees, if a single one does not suffice. Then the engine pulls on the heavy line and drags the whole machine forward; if the gradient be too heavy, another block is used, multiplying the mechanical advantage. Thus the donkey-engine pulls itself forward. It is astonishing to see over what rough country the engine on the sled will advance—down into a creek-bed, up the steep bank, over fallen trees, through the swamp, across ditches. The sled is 40 feet long and 8 feet wide. The 'donkey' also tows a second sled on which are placed provisions, coal for the engine, tools, extra rope, axes, and other necessaries, the entire load weighing fully 30 tons.

Now you will understand what 'siwash' signifies. When a line is passed around a tree, and not through a block, the line is said to be 'siwashed.' Again, a 'choker' is a short piece of cable at each end such as is wrapped round a tree to hold a block. The engine is called a 'donkey,' and it performs feats worthy of the patient 'burro' who is the friend of the Western prospector. But the 'burro' is not in fashion in the North, he is ill fitted either to traverse the tundra in summer or the snow in winter; moreover, his propensity to lift up his voice in song would not accord with the spirit of the Arctic wild over which a Great Silence broods eternally.

Moose and caribou are plentiful in parts of Alaska. The caribou make an annual migration in a vast herd, which has been seen by several men whom I met. Thus Angus Macdonald told me how in the spring of 1902, at the head of Tombstone creek, he "rode through them for a week." A party of prospectors coming south from the Peel river met the herd, which was moving slowly northward while grazing; these men walked within sight of the caribou for six days. In the spring they go to the Arctic slope of the northern mountains and in the fall they return to British Columbia. Smaller bands of 30 to

TOTEMS OF THE THLINGIT INDIANS.

60 are to be found on the Mackenzie range. When these tens of thousands of caribou passed through the Tanana region in 1906 the miners shot them from the doors of their cabins and thus obtained a winter supply of fresh meat.

Soon after leaving Forty-Mile we saw a bear swimming the Yukon. Of other large wild animals seen in the course of the journey, the whales were the most noteworthy. On the return from Sitka to Juneau, as we stood on the deck of the *Georgia*, which was steaming through Sergius strait, a narrow waterway, we saw the fountain made by a spouting whale. Passing close to him, we noted that he was about 45 feet long,

MOOSE SWIMMING.

and as he sped through the water alongside, rising at intervals to 'sound,' we could see the barnacles clinging to his snout and tail. Every time he sounded he made a whistling noise like the wire-drawn exhaust of a large compressor, the air and water shot upward and the dark bulk dived beneath the water, leaving the big tail fanning the air as he disappeared, to reappear in half a minute. This young leviathan was proceeding slowly, evidently feeding on herring, a shoal of which was swimming through Sergius strait.

The law prohibits the killing of game in spring. This would

be proper enough in most regions, but in the distant portions of Alaska the prohibition is a hardship because the spring is the very time when supplies of meat and foodstuffs of every

THE MOOSE-HUNTER.

kind are apt to run short among prospectors and explorers. It should be legal to kill game for food at any time. At Fairbanks, in winter, moose sells for 30 cents per pound; beef, for 45 to 75 cents. On the other hand, head-hunters, killing animals for decorative use in Chicago and New York, ought to be summarily suppressed.

The Alaskan Indian so much resembles the Japanese, and the distance from the Seward Peninsula to Northeast Siberia is so short—only 36 miles across Bering strait—that a racial origin is readily suggested. The Kuro Siwo or Japanese current touches so close to the shores of Queen Charlotte island that it has been suggested that junks from over the sea may

WHEN WIND HELPED MUSCLE.

have been stranded there in those forgotten centuries when the Japanese built sea-going vessels and traveled afar.

Speaking of Japanese, I am reminded of that Alaskan celebrity Jurio Wadda. He is the great 'musher,' a redoubtable explorer, and an indomitable adventurer, giving that word both its modern meaning and the more honorable older one. In Cornwall the shareholders in mines were called 'adventurers,' that is, they were the persons who shared in the venture; and even to this day the president and shareholders of a historic enterprise are called to annual meeting under the name of "the Governor and Company of Adventurers of England trading into Hudson's Bay." Wadda is not so grand as all that, but he is a good deal of a man. His expedition to Herschel island is proof sufficient. With his dog-team and sled he went from Nome to Cape Prince of Wales, thence round the Arctic coast to Cape Blossom, on to Point Barrow, and so to

MINING ON CHICHAGOFF ISLAND.

Herschel island. A glance at the map will indicate the distance covered. He was 32 days making the last stage of this journey from Point Barrow to the island, and for those 32 days he traveled continuously. A little bell-shaped tent was his only cover; in his team were nine dogs, but they were never unhooked from the tow-line or unharnessed during those 32 days. Then he 'mushed' from Herschel island to a point 90 miles east of the Mackenzie river, trying to find the position of a gold discovery the news of which he had obtained from the natives.

Returning to Herschel island, he started back to Nome. The season was near its close and six days from Point Barrow he found himself without food; he cut the seal lashings on his sled and the seal 'mukluks' on his feet, and fed them to the dogs. Leaving Point Barrow, where food was obtained from the Eskimo, he worked his way as far as Point Hope and then in an *oomiak* or skin-boat he paddled to Cape Blossom, where he caught the steamer *Corwin* for Nome.

When one hears the true stories of the feats of travel over ice and snow accomplished by the unrecorded heroes of these frontier mining camps, the much advertised expeditions of scientific-newspaper enterprise seems hollow shows. If any millionaire wants to see his house-flag placed on the North Pole, I would advise him to engage Wadda, Scotty Allen, Louis Lane, Sport Smith, John Hignes, Hart Humber, and Clarence Hawkins. These are the men to get there; when they return they can engage Peary and Nansen to deliver the necessary lectures and submit to the requisite interviews in their stead.

CHAPTER XII.

SKAGWAY.

Alaska holds much beauty, but also ugliness enough to force a contrast. Skagway or Skaguay, as it is variously spelled, is the scrap-heap of creation. As the steamer turns from Lynn Canal and approaches this terminal point in a voyage of exceeding beauty, the charred forest on the right suggests the devastating hand of man, heretofore so notably absent. Long wharves, ugly cattle-pens, and empty warehouses bespeak the activity of the past. A Sabbath quiet reigns, as if to compensate for the most unholy doings of the boom days. Skagway was then the place to 'take a drink' and 'mush on.' The first was easiest, and most favored by the wayfarer. East of the wretched settlement rises a ridge the face of which is painted with garish advertisements and cabalistic signs. The tide is out; the flats, wide and long, are covered with noisome seaweed through which the creek finds a dreary way. Four long weather-beaten spindle-legged piers, lightly braced, reach from the town to the warehouses, which are clad in corrugated iron. The railroad clings to the base of the east cliff. All the frothy gaiety of Skagway is gone. The town is an ungainly collection of shanties, mostly saloons and gambling houses, now out of business. A few good stores and a cheerful group of offices bespeak steamship and railway transport, and suggest that the traveler at least can get away without loss of time.

'Skag-waugh' is the Indian for cruel wind; the natives were afraid of the icy blast that blew down the canyon from the White Pass. Many a man must have shivered as he landed from the steamer and looked at that cold gray landscape where in the distance the savage peaks beckoned to hidden gold in the unknown wilds of the mysterious North.

Skagway is associated with the doings of Soapy Smith, a desperado who terrorized the trail to the Klondike at the time of the rush, at the end of 1897 and in the succeeding year. He was killed on July 8, 1898. His name was Jefferson Randolph Smith, a man of ordinary education, but possessed of extraordinary cunning and unflinching nerve. At one time he was a newspaper reporter in Georgia, and he had operated in confidence games all over the West, especially in Colorado, becoming prominent at Creede in 1891. His sobriquet originated from a scheme for selling soap; in one of the packages of soap sold by him he placed a $5 bill; this always went to a confederate, but the chance of getting it would excite the cupidity of simpletons, who bought 5 cents worth of soap for 50 cents in the hope of getting 5 dollars. Soapy Smith began operations at Skagway in July 1897. He had an organized gang of desperadoes who worked under his direction. They did not attack well known men, citizens of the town, or anyone capable of prosecuting them in case of assault. Their depredations were restricted to 'tenderfeet,' to greenhorns, to those who were friendless and alone. Violence was avoided, as far as possible; the plan usually was to excite cupidity by a 'shell' game, by roulette wheels that were mechanically 'fixed,' by picking pockets, by inciting a fracas or 'rough house,' and by making their victim drunk.

For instance, a man named Stuart, who was the first to come out of Dawson in the spring of 1898, carried $2600 in gold 'dust.' He placed his gold in the safe of a merchant at Skagway and then went to Soapy's saloon. There he entered into conversation with Tripp, one of Soapy's men, the leader himself never appearing on the scene. Tripp played the part of a newcomer, and told Stuart that he had not seen any Klondike gold; would Stuart show some of it to him? So Stuart went to get the sack of gold, and on his return to the saloon a play was enacted for his benefit. Some other men were gambling with a stupid fellow, who was losing money right and left. This was brought to the notice of Stuart and he was

EVENING LIGHT ON LYNN CANAL.

persuaded by the confederates that a chance 'to make easy money' was offered. Thus Stuart was induced to make a small bet, necessitating the opening of his 'poke' or sack. Then, somehow—but no one knew why—there was a rumpus and a fight; Stuart was elbowed out of the back door, and when the excitement was over he hastened back to the saloon only to find that everyone had decamped, and with them his 'dust.' Tripp and others who were present swore to the deputy marshall that Stuart had lost his money by betting.

This incident led people to realize that unless Smith was driven out of Skagway all the trade incidental to the stampede would go to Dyea or to St. Michael. The robbery occurred at 11 o'clock on the morning of July 8, 1898. A mass meeting was called for 8 o'clock that night at Sylvester's store, and the attendance being too large for the store, the meeting was adjourned to the wharf at the foot of State street. It was broad daylight. Soapy Smith and a number of his fellow ruffians went toward the place of meeting in the hope of creating a disturbance and dispersing the crowd before it had determined upon a plan of action. Frank H. Reid, a civil engineer, and two or three others had been appointed to guard the approach to the wharf and prevent any of Smith's men from getting into the meeting. When Smith approached Reid, the latter said: "You can't go down there, Smith." To which the latter replied: "Damn you, Reid, you have been at the bottom of all my troubles. If I had got rid of you three months ago, I would never have had this trouble." He then started to club Reid with his rifle, but Reid caught hold of the rifle with his left hand and pulled a revolver with his right. The revolver failed to go off and Smith managed to point the rifle downward, pulling the trigger, and wounding Reid fatally in the groin. But before he fell Reid also managed to pull his trigger a second time and shot Smith through the heart. Smith had exclaimed a second earlier: "Don't shoot!" He was killed instantly. In his pocket was found a note reading: "The crowd is angry, if you want to do anything, do it quick." This was signed with the initial *S*, and was identified as the handwriting of W. F. Saportas, a correspondent of the New York *World*,

EVENING.

Photograph by E. W. Merrill, Sitka.

Published by Permission.

well fitted to represent an infamous newspaper—the predecessor of the yellow journals that disgrace America today. Reid lived for ten days. A monument to his memory was erected by the people of the town and on it was engraved: "He died for the honor of Skagway."

The rest of Smith's gang took to the hills. Most of them surrendered through stress of starvation. Among these was Tripp. The committee of citizens ordered 35 persons out of town, including Dr. J. Allan Hornsby, editor of the *Daily Alaskan*, also Saportas, sundry barkeepers, thugs, and so forth; 18 were held for action by the Grand Jury; of these 9 were sentenced to the penitentiary, on evidence and confession. It is only just to add that Dr. Hornsby protested his innocence from the beginning and after the excitement had subsided he returned to Skagway, becoming employed as a physician in the hospital service of the White Pass & Yukon Route. This indicates the favorable opinion of the railroad officials.

LORING; A FISHING VILLAGE.

CHAPTER XIII.

THE STAMPEDE TO DAWSON.

Skagway is the portal of the Yukon. From the wharf the rails of the White Pass & Yukon River route lead up the valley and over the White Pass, thence down the watershed of the Yukon to White Horse, where steamers start for Dawson. The summit of the pass is 20 miles from Lynn Canal and 2886 ft. above tide-water; it is 111 miles by rail from Skagway to White Horse, and 460 miles by the river from White Horse to Dawson.

At Skagway the traveler hears much concerning the great stampede to the Klondike, and, if observant, he will see many mementos of that remarkable episode. I went by train to the summit of the White Pass, and then walked down the old trail, so as to obtain a nearer view of the path over which the gold-seekers trudged at the time of the big rush.

The railroad and the trail both ascend the gravel-strewn valley of the Skagway river; when the latter forks, the trail takes the west branch, while the railroad makes a big loop up the east branch before re-joining the path of the Klondikers. The old trail is partly obliterated by the Brackett road. At the beginning of the rush, in the winter of 1897-98, George A. Brackett built a wagon-road as far as White Pass City, 10 miles from Skagway. The toll was one cent per pound. By filing his maps in the Recorder's office he secured the right to charge toll, but it was not collected without an occasional fight. Brackett built two or three gates along the route, hoping to control the migration, but the packers would combine and rush these gates. Thereupon, he employed armed guards, and some shooting ensued. It was claimed that Brackett blocked the old trail at the foot of the canyon above White

Pass City; this caused bitter feeling, but it did not last long, for the railroad company bought his rights. Grading began in May 1898, and the railroad was built to the summit by January 20, 1899; to Lake Bennett on July 6, 1899; and to White Horse on July 31, 1900.

On the west side of the creek a couple of log-cabins mark former outposts of Soapy Smith. It is estimated that from 75 to 80 men were murdered by this desperado and his gang, the victims being thrown into the icy waters of the estuary. At the 12-mile post a splendid cascade is seen in a canyon opposite. Here the railroad begins a sweeping curve to the station called Glacier. The brown pile of débris just above the track marks the moraine of a glacier that reaches from the cirque on the right. From this point the railroad is cut into the face of a precipitous slope; approaching a tunnel the track crosses a high wooden trestle from which a splendid view is obtained of snowy mountains, then the vista framed by the walls of the canyon, and a glimpse of Lynn Canal at the far end. Near the head of the White Pass the train swings round a curve, the track passing over a high steel trestle with concrete piers. The old trail can be seen threading the bush-covered slope on the other side of the gully.

The best outlook is from Inspiration Point, just beyond the big curve at Glacier. The view is down the canyon to Lynn Canal. On the right is the rough and rocky shoulder of a mountain, bare save for moss, being above timber-line; on the left is the straggling upper limit of forest growth, then come glaciated granite bosses, bearing patches of snow, which merge into snowfields and mist-covered summits. Looking down the canyon, the sunlight plays on the bright verdure of the Alaskan highlands, the stream appears as a broken silver thread amid brush and rocks, the dark perspective of the valley leads to a sunlit space of water where Lynn Canal, like an enchanted lake, sleeps under the frowning ramparts of the coast range. Beyond is mist, broken by shafts of light, and the cold breath of an air more akin to the bitter tragedy of fact than the warm romance in which the search for gold is wrapped.

ON THE WHITE PASS RAILROAD.

LOOKING DOWN THE WHITE PASS.

I left the train at the Summit station and followed the old trail down the canyon. This was the highway of those excited Argonauts who scrambled toward the Pactolus of the Yukon in the winter of 1897 and the spring of 1898. The trail is not steep, it requires none of the skill of a mountaineer; indeed, the ascent from Skagway to the summit of the White Pass can be described as a good long walk, with a rise of less than 3000 feet in 16 miles. The other pass, at Dyea, which is in the next valley to the west, was much steeper, especially close to the summit.

The trail descends gently to the meeting of two rivulets, where, on a small flat, was the site of White Pass City. This served as a relay station; firewood and water were handy, and being at the timber-line, it was the last place for a convenient camp. Ten years ago 1500 to 2000 people congregated at White Pass City in tents, log-cabins, shanties, and improvised shelters. Drinking, dancing, and gambling shamed the night and mocked the day. Many a young and healthy man succumbed to perils more trying than fatigue and more deadly than the snow. It was the mockery of the romance of mining, a sordid debauchery, an unveiled licentiousness, the procurers of which were Soapy Smith and his gang of ruffians. Not much remains of White Pass City; half a dozen log-cabins and the frames of a dozen unsubstantial structures bespeak the wreckage of a frontier settlement. The largest of the shaky buildings was a dance-hall; it is now invaded by alder bushes, and alongside the doorway the humble gooseberry grows confidently. The wreck of an old sled is wreathed in blossoms of elderberry. In front of what evidently was a store, a pair of scales and some bottles lie untidily. A delapidated dwelling, with the sign 'Hotel' hanging loosely, suggests the mob of adventurers and harlots that gathered there not long ago. And yet the spot has more than ordinary beauty. Kingsley spoke of "ancient and holy things" that "fade to the earth again"; fortunately, the unholy things decay even more rapidly. White Pass City and its inhabitants are gone, leaving few traces, insufficient to soil the face of Nature. It is a picturesque spot where several cascades meet joyously; the confident curve of

CROSSING THE CHILKOOT PASS DURING THE KLONDIKE RUSH.

the railroad belts the hillslope in front; and far overhead noble peaks look down in eternal calm; the air is perfumed with blossom, the murmur of the stream is soothing, the sunlight suffuses the lush grass. Man's unrest is petty indeed in contrast to Nature's imperturbability.

The motto of Alaska is, 'Mush on!' In the North that is what the mother says to her child, the man to his dogs, the barkeeper to the drunken loafer; that is what destiny says

WHITE PASS CITY. ON THE TRAIL TO THE KLONDIKE.

to Alaska. It was the watchword of the stampede that startled civilization when the gold began to come from the Klondike diggings. To 'mush' is to walk; the word is derived from the French *marcher* and was brought into the Northwest by the *coureurs de bois* and the *voyageurs* from Quebec and New Orleans. The crowd that crossed the passes during the excitement of 1898 were 'mushers,' not mountaineers; they were gold-hunters from the cities, not prospectors; they were 'cheechakos' of the greenest kind. Hence their troubles. If the unholy pilgrim-

THE RUSH TO THE KLONDIKE IN THE SPRING OF 1898.

age to Dawson was marked by the horrors of death and privation, it was due largely to the inexperience of the pilgrims, and also to the exaggerations of the scribes who snatched an ephemeral fame by misinforming a credulous public. For instance, Harry De Windt crossed the Chilkoot pass in 1896 on his way—theatrically speaking—"from New York for Paris by land," and in the inevitable book he submits a picture of the pass by which he crossed the range; it is represented as a precipice, having a slope of 65 to 80°. A man cannot walk up a slope of 40°. This highly imaginative illustration* based upon the memories, and exaggerations, of an irresponsible author, affords a good example of the way in which the passage over the coast range was made to appear a tremendous feat of mountaineering. As a matter of fact all of the passes were ascended by men ignorant of mountaineering and loaded with heavy packs; even the Chilkoot has been crossed by cattle and horses; any one of the ascents would constitute a pleasant excursion for men accustomed to the mountains and unburdened with the supplies needed by the prospector on his way to diggings five or six hundred miles distant. The horrors of the White Pass and the Dyea trails were due to the unintelligent scramble of a mob eager to reach the scene of gold discovery. Men from the cities, unused to open-air life, unaccustomed to packing, wholly ignorant of how to take care of themselves, in a frenzy to reach Eldorado, were bound to get into trouble on a rough trail crowded by others like unto themselves. The 'rush' was composed largely of people unfitted by physique or temperament for the hardships of the frontier. It was a mob of inexperienced men; there was no directing head, no organization; if properly organized under experienced leaders, the whole of the feverish migration

*Wrong ideas concerning the steepness of the ascent along the trails are induced through the foreshortening of the line of slope and also by views of declivities, the angle of which is exaggerated by the cutting of the photographic print so that a false base is made. Men walking up a gentle slope can thus be made to appear climbing a precipice.

might have been effected with a fraction of the labor spent and the hardships endured.

During the winter of 1897 not less than 33,000 men and women passed through Skagway on their way to Dawson over the trails of Dyea and the White Pass. Owing to their inability to transfer their outfit across the range after the snow had fallen, thousands of men were stalled at Skagway, Dyea,

THE STAMPEDE. ON THE CHILKOOT PASS.

and White Pass City. These small settlements became badly congested. The conditions of living were wretched; dissipation, poor food, excitement, and inadequate clothing combined to decimate the mad throng of gold-seekers. In April 1898, 42 were killed by a snowslide on the Dyea trail. During the winter 46 died of spinal meningitis, due to over-exertion and exposure. Many young men from decent homes were victimized; they found a "wide open town," with saloons, dancehalls, and gambling-dens at high tide; they easily went wrong, spending their stake-money. These never reached Eldorado, although many crossed that far range from which no traveler returns.

The difficulties of the Klondiker arose from the need of carrying supplies on his back. The first and last method of transport is human porterage or 'packing.' In a new country like Alaska, where vehicular traffic does not exist, where railways are rarely available, where roads are scarce and trails are poor, the load on a man's back often represents all his possessions, including provisions and tools. It is true that navigation on the rivers, effected in shallow boats and punts, serves to facilitate exploration, but even where rivers give access to the interior, the end of the journey, with many interruptions, is sure to be marked by a 'portage.'

To all men it is hateful to do the work of a beast of burden, and only the 'old hands' will do it with equanimity. The 'packing' killed newcomers at the time of the big rush. But it is a truism that men will get used to almost anything. The 'old timers' spent all summer packing 'grub' for the winter; for example, Joe Barrette told me how he went up the Sixty-Mile river in '93; he went to Miller creek, which is 125 miles from the Yukon, on foot, prospecting on his way; he carried an average load of 100 pounds and made 20 miles per day. Some men carry 125 to 130 pounds habitually. Barrette and his partner made $300 apiece by 'rocking' the river bars on their way to Miller creek. Again, John Flygar, now at Fairbanks, related how he hauled 1000 pounds on a sled from Fort Yukon to the Birch Creek diggings, completing the 150 miles in 20 days. He did it in three relays; thus he was compelled to walk the entire distance five times, his sled being loaded three times, and empty twice. In consequence he tramped 750 miles in the 20 days. This meant from 10 to 11 hours of 'mushing.' The road was well traveled, having been used for hauling during the previous winter. This was in April 1898. In 1899 many men walked down the river from Dawson to Nome, a distance of 1117 miles. In 1898 the Canadian government passed a bill to prevent ill-equipped persons from entering the

Yukon Territory; by this law each person must have 1100 pounds of supplies, or more than enough for one year. Two pounds per day is deemed an adequate ration; one pound of meat and one pound of bread.

At the time of the rush the outfit of the Klondiker averaged nearly 2000 pounds, or one ton. Some of them packed 50 to 100 pounds at a time, making numerous trips; others pulled a sled, carrying 200 pounds. With a harness over his shoulders, tugging a sled heavily loaded, with eyes bulging out, sweating,

THE LINE OF STAMPEDERS.

swearing, excited, the 'musher' would advance a few miles and deposit his load. While he returned for more, his partner stood on guard. Usually the Klondikers worked in parties of two or three, taking weeks or even months before they reached the summit of the pass. After a day of unremitting feverish toil, these men would come back either to a green log-cabin or a flimsy tent. This was dangerous, although it was not unusually cold that winter, the worst being 5° below in February 1898. Nevertheless, the incoming 'tenderfeet' suffered severely, while the seasoned 'sour-doughs' returning from Dawson were 'mushing' with a temperature of -50°, sleeping in tents, and doing nicely. The trails in the valley were mired by the constant traffic and a single path in the snow limited progress to the speed of the slowest in the long file of trudging humanity. No man who now travels up the White Pass in a comfortable railroad train can have any idea of the hardships endured unless he has seen something of the kind in other gold-rushes.

One young fellow 'packed' the 1500 to 2000 pounds constituting his outfit from Dyea over the Chilkoot pass to Lake Lindeman, where he built a boat. Going down the rapids to Lake Bennett, he ran against a rock and lost everything. Thereupon he walked back to Skagway and procured another outfit which he carried, as before, over the pass. Having built another boat, he descended the rapids, struck the very same rock, and was wrecked. Going ashore, he blew out his brains.

A more cheerful tale is that of a Klondiker who bought a newspaper at Lake Bennett. This newspaper contained an account of the naval battle at Manila. On arrival at Dawson the paper was sold for 10 ounces of gold, equivalent to $160, and that seemed an excellent trade, for it had cost only 50 cents. The last purchaser immediately rented a hall and charged $1 admission to those who came to hear him read the description of Dewey's great victory; he cleared "better than" a thousand dollars.

Mules and horses were used in packing. Since the men took so little care of themselves, it was unlikely that they would show any consideration for animals; and they did not. The death-rate among the horses was frightful; it is estimated that 2500 of them died on this side of the summit during the fall of 1897. Angelo Heilprin, who crossed in July 1898, records counting more than a thousand rotting carcasses, the stench from which made travel over the White Pass unbearable.

The Indians helped the gold-seekers. These wretched-looking natives had made a business of packing long before the rush. J. E. Spurr, who crossed the Chilkoot pass in 1896, paid the Indians 7 cents per pound for carrying his outfit, and he makes note concerning the toughness of the Chilkoots, who carried from 125 to 160 pounds apiece over the rough trail a distance of 13 miles and yet at the end of the carry only rested an hour before returning to Sheep Camp. On the Dyea trail the Indians charged 10 to 15 cents per pound for packing the 50 miles. The ascent to the pass is gentle except on the immedi-

ON THE SUMMIT OF THE CHILKOOT PASS.

ate approach to the summit, where, at the Stone House (a big boulder beside the trail), it is very steep for about a mile.

> "So steep the hill the leg was fain,
> Assistance from the hand to gain."

An Indian buck would pack 100 to 140 pounds; a squaw 80 to 100; girls and boys, from 25 to 65 pounds apiece. The portage from Dyea to the foot of Lake Lindeman was done in 12 to 15 hours. On the steep pitch the pack is divided in two. White men have been known to carry 150 pounds, with a pick or shovel extra, as 'trimmings.'

During the rush one man cut steps in the snow for 150 feet

near the top of the Dyea trail. He and a hired man worked for one night only. They collected a voluntary toll from the 'mushers' and made from $80 to $100 per day for about six weeks. Then this enterprising individual "went on a tear," otherwise a drunken debauch, and during his absence another man annexed the steps, together with such rights as he could enforce on those who came that way.

Similar stories are common. Everyone was in a hurry; anything that facilitated progress was liberally compensated. Thus, a man landed at Skagway without a dollar; another man brought 12 horses; the latter "got cold feet," that is, lost his courage to go forward; the former had no capital but pluck, so he borrowed money from a saloon-keeper, bought the horses at a bargain and began 'packing,' namely, the transport of baggage and supplies for the horde of gold-seekers. He made $300,000. And then he also "went on a tear," degraded himself with drink and other debauchery, and blew out his brains. An enterprising fellow threw a log across the stream and charged 50 cents for a dry crossing; the 'mushers' were glad to pay the toll rather than get wet by fording the creek.

One of the 'old timers'—how quickly men and events become old on the rapidly shifting frontier!—told me that he packed 850 pounds over the Dyea trail. On his last trip he carried 147 pounds of bacon. His expression bespoke fatigue, but his face lit up: "I had two men packing money for me when I came out." "Money" meant gold 'dust.' Within a year he was able to retrace his steps down that same trail, this time with a fortune. He was one of the few that kept their heads and returned home with a competence for life.

That is the sunny side of the story: a few level-headed fellows made money rapidly and returned home with enough capital to buy a farm, a business, or a home. Without loss of health or self-respect—on the contrary, the hardier for their experience, both physical and moral, they obtained a new start in the careers open to strong men. The romance of the rush was not with the frenzied 'mushers,' the greedy harlots, or the drunken desperadoes who figure in the tales of the period, but with the quiet strong men who greatly endured and nobly over-

CAMP OF KLONDIKERS ON LAKE LINDEMAN IN MAY, 1898.

came the trials of an unaccustomed life and returned home to be the leaders of a peaceful community.

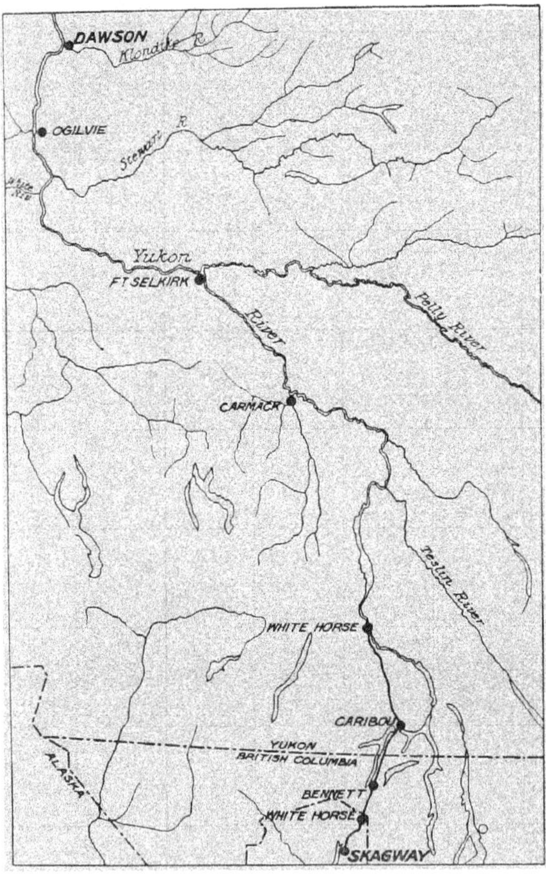

PART OF THE YUKON TERRITORY, CANADA.

CHAPTER XIV.

ON THE WHITE PASS RAILWAY.

Leaving Skagway at 9:30 a.m. the train reaches the summit of the White Pass at 12:20. The distance is 20½ miles, but the ascent is performed leisurely. Here runs the international boundary between Alaska and British Columbia. One of the survey monuments can be seen on the hillside just above the snowshed; it is a cement pillar, on the southern face of which is inscribed UNITED STATES; and on the opposite side, CANADA.

The mountain surface at the top of the White Pass is bleak and bare, for in the past it has been subjected to tremendous glaciation and during the winter it is now swept by terrific storms. Three miles beyond the Pass the traveler on the train gets his first glimpse of Summit lake, one of a chain of milky emerald pools down which the Klondikers floated in the summer of 1898. The lowest of the three lakes, called Shallow, was so true to its name that the gold-hunter left the uncertain waterway for the mossy bank, preferring to follow the corduroy trail, now fallen into decay. A little farther, the railroad traverses an intensely glaciated tract, an area of inundated *roches moutonnés,* in which is Mud lake, bordered on two sides by a moraine sufficiently regular in outline to be mistaken for a railway embankment.

Log Cabin and Tagish Post are associated with memories of boundary disputes; for the Canadians thought the line ran first through one of these points, and then the other. The Northwest Mounted Police collected the duty—25 to 30 per cent—but they had trouble when the boundary post was moved from Log Cabin to Summit, the packers claiming that they were in American territory.

The Klondike trail crosses the line of the railroad several times between Log Cabin and Lake Bennett, and runs parallel most of the way. Those who went over the Dyea trail at the time of the rush came by way of Lake Lindeman, which is separated from Lake Bennett by a short run of rapids. At 1:25 p.m. Lake Lindeman is passed and, looking back, a notch in the range indicates the pass. On a bleak ridge to the left is a cemetery, where lie 28 victims out of some 20,000 people who, in the spring of 1908, camped at the head of Lake Bennett while awaiting the breaking of the ice. All of them had tramped during the winter over the passes from Skagway and Dyea.

At Lake Bennett the train stopped long enough to permit a short walk. By going over the shoulder of a hillock to the west, the traveler obtains a full view of the ruins of the settlement that marked the so-called head of navigation at the time of the Klondike excitement. Skeletons of shanties and a weatherbeaten wharf, the hulk of a boat, and the wreck of a wheelbarrow, some rotting sleds and rusting cans, old boots and bottles galore—that is all that survives.

> "Man marks the earth with ruin,
> His control stops with the shore."

The lake bears an impassive face, and its waters lap the beach as gently as before the "alarums and excursions" of the stampede. The faint murmur of the rapids is borne upon the mountain air like the sound of the surf heard far inland; the wind rustles the brush growing amid the litter left by the gold-seekers, and above the sordid disarray of old boots and empty bottles a meadow lark carols gaily.

The only structure surviving in anything like decent order is the church, built in 1899, but even this suggestion of morality amid sin and canned vegetables appears old and sightless, for the windows are boarded and the bell dismantled. On the west shore of the inlet at the head of the lake are the charred

ON THE WHITE PASS RAILROAD.

remnants of Mike King's sawmill, an establishment that did great service at the price of $80 per thousand, "and upward," as hotel men say. Eight years ago from 10 to 15 steamboats churned the waters of Lake Bennett, and countless vessels of every kind and shape bore the Argonauts toward the golden sands of the Klondike.

The railroad* follows the east shore of the lake, which is 26 miles long. As a lower altitude is reached, vegetation becomes more assertive. On pretty little beaches the force of the wind is indicated by the wrecks of several boats, some of which were lined with canvas and others with sheets of galvanized iron. Approaching the West Arm, the expanse of still water spreads under the shadow of a conical mountain, which, with snow and cloud, is reflected faithfully in the unruffled face of the lake. On the West Arm was another sawmill, now the residence of a solitary hunter and trapper. Near the north end of the lake, the opposite slope exhibits two former shores of recent origin, for they show but little erosion. The railroad gets gravel for ballast from one of these old beaches. At the outlet of Lake Bennett, the train crosses a drawbridge and enters the settlement of Caribou.

Caribou is the point of departure for the Atlin district. The town stands upon a sandy soil recalling the mining camps of southern Nevada. Beyond Caribou the railroad traverses a forest of scrubby growth, devastated by fire and hideous to the sight. It is a sandy tract, formerly the bed of Lake Bennett. The prospect improves and begins to look like northern Michigan, as the railroad follows the course of a meandering stream —the Watson river—which has cut deeply into the sandy gravel of the old lake-bed. Then Minto is reached, 81 miles from Skagway.

---

*Our impressions of the scenery along the White Pass & Yukon route were of the best, for our point of view was delightful. Mr. Scott Turner and I traveled with Mr. A. L. Berdoe, the general manager, and Mr. V. I. Hahn, the superintendent of the road, in their private car. Thus we became indebted not only for adequate nourishment but for information prompted by the sights viewed from the rear of the train.

ON THE SHORE OF LAKE BENNETT.

The Watson has dwindled to a series of pools, and the scenery resembles the 'bad lands' of North Dakota. Old beaches are in evidence and deep narrow gullies, cut in the sand, suggest unusual erosion. Our perplexity is removed by an explanation from Mr. Berdoe: The railroad company wanted to lower the level of Lake Lewis in order to use the level shore for the grade and shorten the line by several thousand feet. It was planned to lower the water about seven feet; a ditch was dug 10 feet deep and 5 feet wide for a length of 350 feet. The water ran

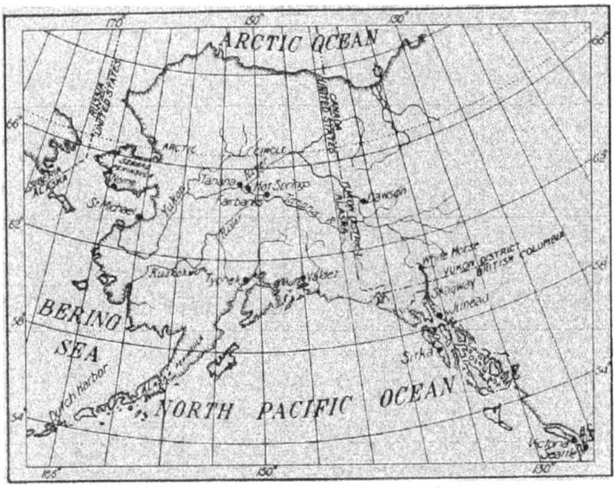

ALASKA.

gently for six hours and then cut its way with increasing force, until it rushed violently along a channel 500 feet wide. Lake Lewis was lowered 83 feet and was emptied into Lake Bennett. Thus geologic action was accelerated and the scenery was spoiled, so that now the weird expanse of sand and reeds looks a fitting habitat for a dinotherium or a glyptodon.

Leaving the shrunken waters of Lake Lewis, the landscape becomes less abnormal; pleasant bits of water with ducks swimming on their surface and grayling underneath, with a fringe of rushes along the shore, and pine, cottonwood, and spruce for a background. But the outlook is soon obscured by smoke,

LAKE LEWIS AND THE YUKON RAILWAY.

due to careless 'mushers,' who fail to extinguish their campfires and allow the wind to blow cinders into the dry grass between the trees. This year an unusual number of men have tramped from Skagway to White Horse, on their way to Dawson. The increase is imputed to bad times in the States.

Approaching Wigan, 105 miles from Skagway, the Yukon is seen dashing through Miles canyon. At 5:30 p.m. we reached White Horse.

REMAINS OF THE KLONDIKE RUSH, ON LAKE BENNETT, 1908.

CHAPTER XV.

WHITE HORSE.

White Horse; the origin of the name is obvious: the crests of the waves in the rapids where the Yukon flings its white mane to the breeze in a mad gallop over the rocks. Owing to the dangers of the rapids, the Klondikers found it advisable to make a 'portage'; thereupon, a tramway was built along the bank, and a ropeway was stretched across the river, and thus there was ample reason for the rapid growth of a settlement, which served as a depot on the way to the diggings. At White Horse the Yukon widens and steamboat traffic begins. Here is the terminus of the White Pass & Yukon railroad and the headquarters of the winter stage service for Dawson. The casual visitor will gather the impression of a populous burg from the bustle and activity of its inhabitants and hardly realize that the summer population numbers no more than about 500. For these things are relative, and after you have traveled for several days and have seen only two men and a dog, a settlement like White Horse is impressive.

The town is situated upon a flat bounded on the west by sandy bluffs 100 to 150 feet high; the main street runs along the river bank, where the Lewes or Yukon sweeps by with a 5 to 6-knot current. The railroad terminus is marked by long gray warehouses made of corrugated iron, facing the wharves, to which steamboats are moored. One of these is the *White Horse* on which we expect to sail to Dawson; the other is the *Prospector*. A little farther down the river are the 'ways,' on which steamboats are raised from, and lowered to, the water. The accompaning photograph shows the arrangement. The 'ways' are timbers (12 by 12 inches) arranged parallel, the boat sliding over them in a direction at right angles to their length. These constitute the 'standing ways' and upon them is placed the 'slide,' which is a 9 by 12-inch timber, with 'slippers' 3 inches deep and 4 inches thick, to keep the 'slide' in place. The surface of the incline is greased with tallow, on which, after it has hardened, dogfish oil is smeared. The gradient being about one inch per foot allows the flat-bottomed steamboats to be lowered by gravity, or to be hauled onto the 'ways' by four capstans, with four tackles and two horses to each. The steamboats are propelled by a paddle-wheel at the stern, and are made of a double layer of 3-inch plank attached to a frame constructed of 3 by 6-inch material. Two large boats, the *Yukoner* and the *Canadian*, stood high and dry on the ways. The *Yukoner* is 173 feet long and 32 feet beam; she has not run on the river since 1903, for her draft is 3 feet, which is too much for the shallows of the upper Yukon. Most of the boats draw 18 inches to 2½ feet. By peering underneath her flat keel we could see where she had scraped over the bars. Sheets of galvanized iron attached to the sides serve as a protection against floating ice.

The water-front of White Horse is eloquent of the mining operations of the interior; here are seen the supplies and machinery consigned to Dawson. At the time we were there, in July 1908, rows of huge pipe bespoke the construction of the big ditch and water system of the Yukon Gold Co. Pipe from 45 to 54 inches in diameter and ⅝ inch thick (some of it manufactured in Germany), suggested a colossal undertaking and a lordly expenditure, while the beginnings of agriculture were indicated by a plough and a mowing machine.

On our arrival we found a comfortable hotel, and after 'supper' we wandered along the river front. It was 8 o'clock, but, of course, broad daylight, for we were now in latitude 61° north. A carpenter at work afforded a source of information. He stated that he was constructing a boat to hold four men and their provisions; the cost of the boat was $20 and represented one day's labor on his part, with 140 feet of dressed lumber worth from $55 to $75 per thousand. If the men traveled by steamer the second-class fare was $30, while the boat

WHITE HORSE, YUKON TERRITORY.

cost $5 apiece for four men and the grub about $2.50 per man. According to the regulations enforced by the Northwest Mounted Police all travelers must give their names, and upon their boat a number is placed, so that the Police may keep account of accidents, in case boats fail to reach Dawson, and trace criminals.

Here we obtained an inkling of the supervision exercised by the Northwest Mounted Police, the finest body of men in the world engaged in such service. They are the best embodiment of authority and to them is accorded the highest respect in the enforcement of law. No traveler in northwestern Canada, whatever his nationality, can fail to be struck by the contrast between these men and the usual type of inefficient municipal policemen. When calling upon the officer in command at the barracks I ventured to suggest that the respect for law in the Northwest was due to the efficiency of the Mounted Police, but he modestly and truly attributed the result to the prompt enforcement of justice and the absence of political undercurrents.

The Mounted Police was originally modeled after the Irish Mounted Constabulary, and grew out of the necessity of the Northwest when that vast region was surrendered by the Hudson's Bay Company to the Canadian government. The Hudson's Bay Company was organized in 1670 under a charter granted by Charles II to the Bavarian soldier, Prince Rupert, and the friends of the latter. Prince Rupert, the first Governor of the Company, was succeeded by the Duke of York, afterward James II, and by John Churchill, afterward Duke of Marlborough. The only return asked by Charles II for the empire he gave to the Company was "two elks and two black beavers." The beaver was the lure to British Dominion in northwestern America, as the sable was the prize for which the Russians traversed Siberia and invaded the country that is now called Alaska. When the Dominion of Canada was but five years old, it acquired 2,300,000 square miles—a continental area—from the Hudson's Bay Company. The Canadian Pacific Railway was being built and protection was necessary for those engaged in this work and also for settlers. Thus, in September

1873, at Toronto, Colonel George A. French of the Royal Artillery, under the premiership of Sir John Macdonald, organized the Northwest Mounted Police. Originally it was a body of 300 strong. Col. French was made the first Commissioner. Whatever feeling may be entertained against policemen, animosity is rarely felt against disciplined soldiers wearing the King's uniform. Therefore the men were given the red coat, but without any furbelows. As Sir John Macdonald said: "I want as little gold lace and fuss and feathers as possible, not a crack

STEAMBOATS ON THE STOCKS AT WHITE HORSE.

cavalry regiment, but an efficient police force for the rough and ready—particularly ready—enforcement of law and justice." Applicants had to be able to read and write "either the English or the French language." The officers at first were taken from the active militia, but their titles are not of a military character; the chief is called Commissioner, the next in rank are the Superintendents, and then the Inspectors, followed by Sergeants and Corporals, below whom come the main body of Constables. They resemble the *rurales* of Mexico and the *guardia civiles* of Spain. The pay is small. A newly enlisted man gets 60 cents per day, to which is added 5 cents per day annually until his pay reaches $1, with 50 cents extra when serving in the Yukon Territory. Thus the regimental pay ranges from $1.10 to $1.50 per day. Anyone engaged in work outside of the ordinary regimental duties—and nearly every member of the force is thus engaged—earns from 50 cents to $1.50 more. They also receive food, clothing, medical care, and so forth. The policeman of San Francisco gets $120 per month, plus graft.

The officers of the Royal Northwest Mounted Police rank with the militia, and are usually accorded equivalent titles; thus the Commisisoner is called Colonel and the Assistant-Commissioner is a Major. They come from military colleges, from civilian appointments, and from the ranks. An intelligent man wins rapid promotion, one-half of the officers having risen from the ranks. After 20 years of service a pension equal to 40 per cent of pay is allowed and by continuous service the pension is raised to the ratio of 70 per cent. Officers can retire after 25 years service with a pension of 50 per cent, increased to 70 per cent at the end of 35 years of service. The men are above the average in physique, intelligence, and morality; last September only 6 out of 160 applicants were passed. As a rule these Northwest Police are good woodsmen, excellent horsemen, and handy in a canoe. Most of their traveling in the Yukon is done in a light wagon, riding only on the regular trails. In winter they go across country with dogs, and in summer they 'mush.' Owing to appointment by the Dominion Government, this constabularly is independent of local politicians and of changes in public sentiment; it is one of the regulations that members of the force must not show political partisanship.

The Northwest Mounted Police is answerable to the Dominion Government in the person of the Premier of Canada; the criminal laws are the same throughout Canada; therefore there are no complications through extradition. If one of the Police goes seven miles beyond the boundary of the Province in which he is stationed, he gets the local Justice of the Peace

CROSSING THE YUKON IN WINTER.

to back his warrant, but he is not amenable to the local authorities. When in the Yukon Territory a commissioned officer has the powers of a justice of the peace and of a coroner; in effect, he becomes stipendiary magistrate.

A record is kept of everyone entering the Yukon, from arrival to departure, so that the Police can place a hand on them if necessary. For instance, last May three men left White Horse in a boat (No. 113); about 12 miles below Selkirk one of the men tried to murder his two comrades and killed one of them, wounding the other. The murderer started to go down the river but, being unable to account for his partners, he was arrested. By referring to White Horse all the necessary particulars were obtained. In June he was tried and, four month later, sentenced to be hung. The postponement of the hanging was due to the fact that the case had to be sent to Ottawa for revision and allowance had to be made for delays in the mail. Any man going down the river in a small boat is noted by the Police, a description is recorded with a statement whence he came. No murder goes unpunished, as is shown by the fact that while nine murders have been committed in the Yukon since the rush to Dawson, in eight cases the criminal has been hung and in one instance sentenced to imprisonment. Trial is before the Judge of the Supreme Court of the Territory and a jury, but the evidence brought forward by the Mounted Police has been so complete as to compel the juries to find a verdict of Guilty. In the case of robberies, it has been more difficult to get convictions, and the blame is placed on Americans of lax sentiment. The Police enforce all the ordinances of the Territory and the criminal code of Canada.

There can be no reason why, to the already multifarious duties of the Mounted Police, there should be added that of supplying information about mines, yet in the official reports of the officers in command at the various posts are found descriptions of mining operations and conclusions concerning the richness of mines. These are, of course, of no value, save as reflecting local gossip, and the publication of them does harm, in as much as many people are simple enough to believe, for example, that Commissioner Wood's opinion concerning the

White Horse copper deposits is that of one qualified to express a correct view, just as there are people who mentally bow to the views of a bishop on military affairs, or an admiral on gold-dredging. In the report for 1907 Commissioner Wood states that "the success of the camp is now assured." He refers to copper mining in the vicinity of White Horse. I visited the copper deposits, accompanied by Messrs. J. W. Bryant and Scott Turner. It is not necessary to describe them, for this would require technical details; I have done so in a more suitable place.* Suffice it to say that "the success of the camp" is without assurance, although not without hope, if copper rises in price, and if more work and less talk be devoted to these interesting deposits of ore.

On returning from the copper mines we followed the Dawson stage-road, fringed with briar-roses and the bright flowers of a brief summer season. After the close of navigation the stage makes one trip weekly over the 330 miles to Dawson, and when the rivers are frozen this service is doubled. During the intermediate period, when the steamers are not running, but the rivers are not completely frozen, the passengers and mail are conveyed across the streams in canoes. The service is adjusted to suit the business available; at the end of March, when people are returning to the 'inside,' two or even three stages leave White Horse each day. The stage-coach is a wagon or a sleigh, according to the condition of the roads; as a sleigh it holds 13 passengers and "the driver breaks the 'hoodoo,' making 14"; when replaced by a wagon, 10 passengers find accommodation. In spring, a team of six horses is required; and in the fall, four, according to the state of the road. When the ground is partly bare, the sleighs are pulled by six horses, otherwise four suffice. The fare is $75 to $125 from White Horse to Dawson, with meals and lodging extra. The company does not run the road-houses, so the passengers pay $1.50 per meal and $2 for a room. When all is serene, the trip consumes five days, so that the road-house expense is about $30. From 700 to 800 people make the journey each winter; of these 500 are going

**Mining and Scientific Press.* December 5, 1908.

'in' and 300 'out,' the difference being due to the preference for the steamboats when coming out and the necessity for getting 'inside' early in the spring, before navigation has been resumed.

In the course of the 330 miles to Dawson, 15 teams are used, permitting three changes of horses every day. Each team travels an average of 22 miles. They are fed compressed hay or oats at 4½ cents per pound. Travelers unite in commending the excellence of this stage service.

The traveler who reaches White Horse will want to see the rapids after which the town is named; but at this point it will be necessary to elucidate the nomenclature of the Yukon and its tributaries. The river that sweeps past White Horse is the Lewes, which along a portion of its length is also called the Thirty Mile; 90 miles below White Horse, the Lewes unites with the Hootalinqua or Teslin river, and it may be questioned which of these is to be regarded as the main stream of the Yukon. According to volume of water discharged, the Lewes is the big brother, and I agree with Angelo Heilprin that it will be simpler, and correct, to regard the Yukon as rising in the small lakes above Lake Lindeman, so that the Lewes, the Thirty Mile, and other local names may be disregarded. The Yukon has its source within 25 miles of the ocean to which it returns after flowing 2300 miles.

The river, five miles above White Horse, cuts through three flows af basaltic lava, and in doing so has made the rapids named Miles, Squaw, and White Horse. On emerging from the narrow gorge through the basalt the Yukon swings out into the soft beds of drift and sand, making broad shallows. During flood time it is possible to shoot the rapids with reasonable safety; as the river dwindles, navigation becomes more dangerous. Many men tried to shoot the rapids during the 'rush,' from 1897 to 1899, and of these fully 200 were drowned. At the end of last June two of the pilots on the upper Yukon were lost in the White Horse rapids. When the water is low the river tumbles over the rocks in the centre of the channel, creating a chute as destructive as a waterfall. Both of the pilots mentioned were drowned, the body of one of them being recovered 14 miles down-stream. He was buried on the day of our arrival at White Horse.

To obviate the perils of the angry river, a portage was early adopted. Tramways were built on both banks in the spring and summer of 1898; the one on the west bank was 6½ miles long and was built by John Hepburn, who sold it for $60,000 to the owner of the tram on the east bank. This was 5 miles long and was built by Norman Macaulay. The toll was 3 cents per pound and $25 for boats, but the latter were usually piloted through the rapids for $20. A pilot could make 10 trips per

OLD TRAMWAY AT WHITE HORSE.

day, riding on horseback along the trail from the lower end of the rapids to the head of the canyon, where there was a camp of a dozen temporary dwellings. This was Canyon City. On the completion of the railroad in August 1900, the tramways, of course, became idle and the owners distributed their money in the 'wide open' town of White Horse.

We walked along the abandoned tramway on the west bank into the small grove of alder and cottonwood fringing the river. An old truck and scattered car-wheels suggested the days gone-by. The tramway was a crude affair, the track consisting of rough hewn 4 by 6-inch timbers, laid 3 feet apart. At intervals of 5 to 12 feet cross-pieces or ties were placed. A truck with iron wheels was pulled by a horse. Thus boats and supplies were transferred past the rapids.

While inspecting the old tramway we had our first experience with mosquitoes. Everybody who goes to the North must have his fling at this voracious pest. I am unable to add to the fearful stories that represent Alaska and the Yukon as being in the possession of the powers of Beelzebub. During the three months spent in the country, we were annoyed by mosquitoes on several occasions; once or twice they were so pestilent as to cause me to forego a breakfast or supper in camp, preferring to protect my head under gauze rather than expose myself to their attack while eating or drinking. At White Horse I had a protector made; this consisted of the finest black tulle in cylindrical form, the top provided with an elastic band to attach it to the hat, and the bottom with another band to go around the throat. The base was further covered by a handkerchief; at the level of the mouth was a thin hoop of fish-bone, to keep the veil from coming close to the face, giving a breathing space and preventing mosquitoes from biting through the mesh of the tulle. This proved an ample protection and was readily folded into a package that would go into my coat pocket. Of course, I wore gauntleted gloves. Thus accoutred I would write peacefully in my notebook and observe the sights without the constant irritation of a swarm of insects, both gnats and mosquitoes. At night when in camp I found it well to throw a yard of tulle over my head. In consequence, my memories of Alaska are not clouded by swarms of mosquitoes.* At times when my armor was discarded I realized that the scenery of the country would be beautiful if not shared with so many rapacious and triumphant members of the genus *culex*; but such a mood quickly passed, and I am glad to leave to other travelers the tale of battle with the enemy who comes not singly but in battalions.

*One of the best antidotes against mosquitoes is oil of citronella, which is distilled from a plant cultivated in Ceylon and the Straits Settlements.

CHAPTER XVI.

ON THE UPPER YUKON.

After two pleasant days at White Horse, we went aboard the steamer of the same name on the evening of July 11. Departure was delayed until midnight, while loading 260 tons of machinery, consigned to the Yukon Gold Company, on a barge. This barge was then attached by steel cables to the bow of the *White Horse*, as the accompanying photograph will illustrate. The barge was 106 feet long and 34 feet beam; the steamer was 167 feet long and carried 180 tons of freight; and though the barge was 60 feet shorter, with the same beam, yet it was able to carry half as much again as the steamer behind it. The *White Horse* drew only 4 feet of water and burned about 100 cords on the round trip to Dawson (making 920 miles altogether); of the wood consumed, 75 cords were burned coming up-stream. The fuel cost $5 to $6.50 per cord, the wood-cutter paying the government royalty or 'stumpage' of 50 cents per cord.

At 4 a.m. the steamer stopped at the entrance of Lake Laberge. The channel was silted, as is frequently the case. At the beginning of the season, pilots are sent down to stake the course; but the débris carried by the swift river and deposited at the inlet, forms a shifting bar. Lying in my bunk I could hear the look-out calling the soundings (made with a pole), and occasionally the boat would crunch against the sandy bottom. The sturdy hulls of these steamers find no counterpart in the flimsiness of the upper works; and with the desire to minimize weight there is no attempt to save the weary passenger from the sleep-destroying jangles of the bell-signals and the rattling of the steering-chains.

At 8 o'clock we were in the centre of the lake, advancing slowly. The opposing shores were half a mile and three miles distant, respectively. It was a sunny morning and the air was most stimulating. On the near shore, eastward, the limestone hummocks showed intensely glaciated. The geologist Dawson has referred to these evidences of ice-action and of the sloping surfaces "so smooth that it is difficult to walk over them." Nowhere is the sculpturing hand of the glacier more obvious; several river terraces belt the hillsides far above the present channel of the river, and the fundamental facts of geology are vividly suggested.

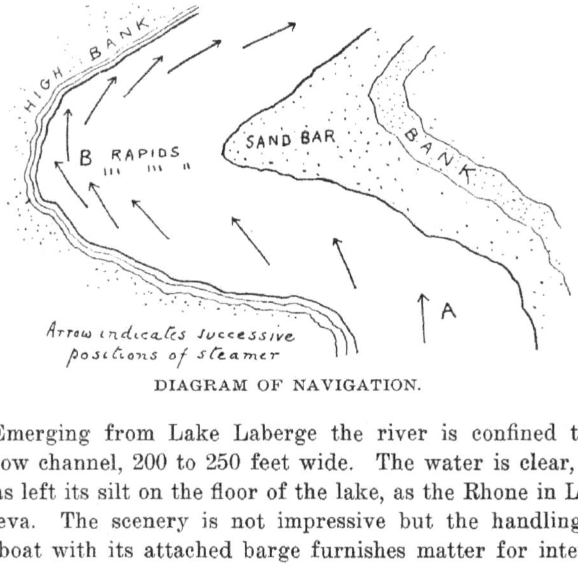

DIAGRAM OF NAVIGATION.

Emerging from Lake Laberge the river is confined to a narrow channel, 200 to 250 feet wide. The water is clear, for it has left its silt on the floor of the lake, as the Rhone in Lake Geneva. The scenery is not impressive but the handling of the boat with its attached barge furnishes matter for interest and comment. The river is crooked and the navigable channel swings from side to side according to the erosion of the banks. The 6 to 7-knot current compels the pilot to be prompt. A bend in the river is characterized by a sandy beach on the inner side of the curve, while across the channel the deep water hugs the steep bank, as the accompanying diagram will explain. The paddle-wheel at the stern acts as a pivot on which the boat turns in obedience to the five rudders under the wheel. Care is taken not to get both bow and stern in the current at the same time, and when the boat has turned into the swift current (at A) the engines are reversed so as to prevent the boat from being carried against the near bank. In making the quick turn

THE BARGE IN FRONT OF STEAMER 'WHITE HORSE', SHOWING METHOD OF ATTACHMENT.

to the right (at B) the boat is run close to the left bank so that the force of the current will swing the bow around; if it fails to do this, then it becomes necessary to back-water in order to give the current time to aid the helmsman. As soon as the turn is made, the signal is given for full speed ahead, thereby prevented the stern from swinging into the shore as the bow comes round. At B the bow of the barge in front of the *White Horse* was within a yard of the shore and to the unsophisticated disaster seemed imminent. An hour later the barge, forming the fore-bow of the steamer, ran aground, and the paddle-wheel was reversed until the current swung the barge out of the sand. A man was sent to examine the hold of the barge to see if she had sprung a leak. The river makes a series of serpentine windings and the pilot needs to be alert in order to clear the stern while protecting the bow of the tandem vessel. The engines are as often in reverse as in forward motion, for the speed of the craft is clearly due to the current, steam-power being used almost solely to gain steerage-way. Often the feeling comes that the boat is grounded, this being caused by vibration from back-paddling, increased by air dragged under the keel by the churning of the paddle-wheel.

Soon, however, interest in the pilotage staled and the sudden changes of direction ceased to attract notice. The boat swept along with the current, through a landscape consisting of scrubby forest, shelving sandy shores, banks of gravel, and a meandering channel. The flutter of a Union Jack gave a touch of color and authority to the cabin marking an outpost of the Mounted Police. Later, a boat came alongside with a couple of men who traded 75 pounds of grayling (a variety of salmon) for flour and sugar. A trail along the east bank marked the telegraph line, the wire of which was rarely visible on account of the foliage. The wreck of the steamer *Domville* suggested possibilities, without intimidating us with the idea of a probability.

At noon, soundings were taken—"Mark Five," was called, indicating a depth of five feet—preparatory to mooring alongside the bank, where a stack of cordwood stood ready. This was loaded by the crew, carrying two or even three logs at a trip, over the gang-plank to the boiler-room. The process is called 'wooding-up'; it was not impressive, and it seemed a reflection upon the coal of the Tantalus mine, which lies a little farther down the river.

Then comes the junction with the Hootalinqua or Teslin,

bringing muddy water into the Yukon. The Hootalinqua is much less sinuous than the Lewes branch of the Yukon; it is also less swift; and on the map it appears as an elongation of Lake Teslin, itself 125 miles in length. At one time it was planned to establish navigation on the Teslin, connecting, by Telegraph creek, with Glenora on the inlet of the Stikine river and thence to Wrangell on the inland sea. The contractors, Mackenzie & Mann, had organized a scheme to this effect, but the failure of the Canadian government to give a subsidy

STEAMER AND BARGE ON THE WAY TO DAWSON.

caused them to desist. Several of the steamers now at White Horse were meant for this service, and the one on which we traveled had been intended to ply on tide-water between Wrangell and Glenora. (See map of southeastern Alaska.)

The next morning I awoke to find our boat alongside the bank, receiving a supply of cordwood. We met the *Victorian*, burning coal from the Tantalus mine. At the junction of the Little Salmon river we passed an Indian village and a station of the Mounted Police. It is the custom to stop small boats and take the names of wayfarers. A boat will drift 60 to 70 miles per day of 10 hours when going down-stream; it is an easy and not unpleasant mode of travel. Up to July 8, in about five weeks, not less than 270 boats, carrying 1200 men, had gone down the river since navigation opened, and 1500 had passed on steamers. In March and April, before the ice broke, 300 men, mostly Slavonians from Treadwell, had walked over the winter trail to Dawson. One of the stages met 68 men tramping, the average time made by these 'mushers' being 11 days from White Horse to Dawson. The road-houses, being placed at intervals of 10 to 20 miles, furnish places for food and rest to this straggling body of immigrants.

Below the Little Salmon a white tent on stilts came into view; this was a 'cache', a term taken from the French-Canadian trappers, signifying a hidden store of provisions. On a platform, six to ten feet above the ground, smoked fish and dried meat are placed so as to be out of the reach of animals, especially of the dogs used for traction in winter.

In the afternoon highly tilted strata appeared on the right bank and black little holes where prospectors had dug into a seam of coal. Then the butte, or knob, above the coal mine at Tantalus came into view, and shortly after we approached the mine itself, marked by coal-bunkers on the left bank and a tramway leading to the entry. At the wharf was a barge loaded with coal for the domestic use of the people at Dawson. A mile down the river is Carmack's trading post, which was an important point before the rush. At 5 o'clock, the Northern day being still young, we approached the celebrated Five Finger rapids and prepared for excitement.

Here four islands of rock divide the Yukon into five streams. As the steamer swept forward under the combined force of her own steam and the current, it was difficult for the passengers to guess through which of the openings the pilot would take her. It was the one nearest the right bank, yet the boat seemed to be heading for the central island. Suddenly the current swung her round into the right-hand channel, and in a moment we were in the midst of the rapids and beyond them into quiet water. The passage was made before anyone had time to develop timidity.

Just above the rapids is a hut and near it a steel rope, 2500 feet long, is anchored; it goes under water and is attached to a ring-bolt on the shore below the rapids. This is used by steamboats that cannot make headway against the current; the cable is passed around the capstan and thus the steamer pulls

THE STEAMER 'WHITE HORSE'.

herself through the rapids. In the photograph some of the crew are seen at work with winch and rope. Before this contrivance was established, boats of small power had great trouble in stemming the current; they would get their boiler-pressure up to 250 pounds and when about to 'buck' the rapids numerous sides of bacon would be stacked ready for feeding into the furnace. If they failed, as they often did the first time, there was nothing to do but to get up steam again and go forward under excessive pressure with the chance of a boiler explosion. During the Klondike rush the small boats and rafts took the opening on the extreme right, and a story is told of two Irishmen, drunk, in a boat, coming through when it was dark. They wanted to know where the Five Finger rapids were and on being informed that they had come through them, one of them replied that he thought the water was "a bit rough". Since then navigation has been improved by blasting some of the rocks that endanger the passage.

Six miles lower the Rink rapids are traversed. They look worse than the Five Finger; as a man who participated in the rush of '98 said: "The white line of foam was calculated to put the fear of God into a man." He hugged the right shore, where a narrow lane of smooth water may be followed.

At 6 o'clock Yukon Crossing was passed, and mail was landed. On the left bank are the road-houses of the stage company, for here the winter trail crosses the river. The Yukon had now grown to a noble stream 250 to 300 yards wide; having conquered every obstruction the river flowed quietly toward the gates of the setting sun. A glorious sunset at 9:30 was succeeded by a moon more lovely than the one Endymion saw.

The next day opened gray, but the vegetation was less stunted, willows appeared, and at rare intervals a patch of green told of a settler's effort to tame Nature to fruitful purpose. At noon we were at Sixty-Mile, where the river of that name enters the Yukon. In the days before Dawson existed the trading posts on the river were named according to their distance from Fort Reliance, formerly an outpost of the Hudson's Bay Company. This was 6 miles below Dawson; even this explanation will not elucidate the nomenclature, for it is 48 miles from Dawson up to Sixty-Mile and it is 52 miles down the river from Dawson to Forty-Mile. The differences are due to early guesses as to distances. At Sixty-Mile Joseph Ladue had a store; he was one of the men made rich and famous during the rush. The old road-house and police station are deserted;

COMING UP THE FIVE FINGER RAPIDS.

only a telegraph office now marks this famous point of departure in Northern exploration. The tributaries of the Sixty-Mile were first worked for gold in 1886, access to the district being obtained by ascending Forty-Mile creek from its confluence with the Yukon and thence across the divide separating Forty from Sixty-Mile. This was the first systematic mining in the Far North. Work is still in progress on Glacier and Miller creeks. Dredging has been started successfully on the lower reaches of Forty-Mile. A government road has been built from the Yukon, opposite Dawson, to Glacier creek, a distance of 73 miles.

We were now approaching Dawson. High hills enclosed the river, brush covered their slopes, and small trees clustered in every protected hollow and on the low-lying flats. On the right was Indian river and Henderson creek, with memories of pioneers and rushes that proved a fizzle. A big hill came into view; it was the famous Dome. The line of the Acklen ditch appeared across the face of the hill and then a big patch of broken rock marking the landslide behind Dawson. The steamer ran alongside the wharf at 4:15 p.m., nearly three days after leaving White Horse. It was a longer trip than usual, for one of the engines was out of order; but that seems a small matter now. We had reached Dawson; that sufficed. Before landing I took off my hat, mentally, to the memory of that energetic observer and capable geologist, one of the scientific pathfinders of Canada, after whom this modern Eldorado was christened—George M. Dawson.

CHAPTER XVII.

DAWSON.

A stroll through Dawson gives an impression of respectability compelled by impoverishment, of the temperance that succeeds dissipation, of the bust after the boom. The town

IN THE ENVIRONS OF DAWSON.

faces the river, and covers a flat just below the confluence of the Klondike with the Yukon. The high ridge, culminating in the Dome, throws a protecting shadow over the straggling settlement. The streets are unpaved black loam, luckily dry just now, and the wooden sidewalks, in places rickety, are weather-beaten but clean. The main street follows the water-front and the wharves are more numerous than required by the diminished traffic. A narrow-gauge railway that goes to the 'creeks,' and three steamers loading or unloading, betoken mining activity not far away. Overhead the tall tower of the ferry and across the broad river the catenary curve of a steel cable mark the manner in which a crossing is effected. Northward are many untenanted buildings, and even the centre of the town bears a bedraggled appearance, indicative of shrunken commerce. During the boom days the population was 50,000; now it is 2000. Dawson looks like a stout man who has grown very thin and yet wears the cloths made for him in his adipose days. Although it has been difficult for Dawson to accommodate itself to straightened circumstances, the adaptation has been effected heroically. The boom has gone, but business remains.

As our steamer hove alongside the wharf there was a big crowd to greet the incoming passengers; the sight was cheerful, but it indicated that the spectators had spare time in plenty. A short walk led to the Regina hotel, a three-story structure sheathed in gray corrugated iron. Hideous it was, but the explanation offered a satisfactory excuse, for the building is made of logs and is covered with iron as a protection against fire. Inside it proved to be an attractive wainscoted hostelry.

On the outskirts, at the foot of the green slope rising to the Dome, are clusters of little log cabins, pretty and picturesque as a stage-setting, with overhanging eaves and flowers. Women in white frocks and a sunny cheery look in the faces of both man and Nature all bespeak the spring following a long winter. In the store windows fruit is offered at fabulous prices; also nugget jewelry,* and furs. Every patch of soil not covered

---

*In a jeweler's store a flat piece of gold, about the size of a man's hand, was on exhibition. This was found in the summer of 1903 at the mouth of French gulch and Eldorado creek; it weighed 86½ ounces, was the largest nugget ever discovered in Canada, and was bought by Clarence J. Berry during my stay at Dawson. Mr. Berry was one of the first men to come with gold from the Klondike in the spring of 1898 and he was one of the successful ones whose story excited the world at that time.

DAWSON AT MIDNIGHT JUNE 20, 1908.

by dwellings is green with new grass and bright flowers. Window-boxes of pansies and mignonette lend a sentimental aspect to the log-cabins. Saloons that are closed and the Mounted Police are suggestive of cause and effect. Nothing rowdy or exuberant survives. Dawson looks like a nicely washed young man of subdued demeanor who gives a hint that he has been extremely gay.

Returning to the hotel, the idea of dinner looms large. Soon we are sitting in a dark low wainscoted room, old-fashioned and cozy, beside an open window, affording an outlook upon green hillsides. The dinner is given a rare distinction by the introduction of a steak that looks like beef and tastes like venison, but is incomparably superior to either. It is young moose, delicious and savory; preceded by fresh salmon from the Yukon and followed by a good cup of coffee, plus the first cigar for a week, it affords keen satisfaction. I shall remember the occasion for many a year. Weary of the steamer and the rough people on board, tired of canned vegetables and a dirty service, it was good to be on land, sitting at a clean table, in congenial company. The air entering the window was of exquisite purity—you must go North to understand that this is no exaggeration—the outlook was pleasant and restful, and in the hall of the hotel a band played 'Sognando.' It was not perfect music, but it served an excellent purpose, linking the cheerful present to the happy past. *Sufficit.*

Next day we saw a little more of the town, preparatory to delivering letters of introduction. Dawson has several large buildings, being the capital of the Yukon Territory. The Administration building, the Court House, and the Commissioner's office are all wooden framed structures painted a neat gray; sensible, and sightly. A Carnegie library enclosed in tin-plate, simulating stone, serves as an ugly monument to a great rebater desirous of notoriety. The barracks of the Mounted Police and the jail or 'skookum house' are built of uncovered logs. Ten years ago the flat and the lower hill-slopes were covered with trees, all of which were rapidly cut down as necessity required.

Meeting a former Dawsonian, who had been a fellow-passenger on the *White Horse*, he began to bewail the quietness of the town: Oh! for the 'old days' when bars and dancing saloons stretched for the depth of a whole block, from street to street; when the crowd within was so dense that it took half an hour to go from the front to the back door; in those brave times there were more people in one block than were now to be seen in the whole town. This *laudator temporis acti* became *splendide mendax*, explaining in strong vernacular how *sic transit gloria mundi* and the deep meaning of *hinc illae lacrimae*—all of which

THE REGINA HOTEL, DAWSON.

goes to show that Latin quotations are a snare and delusion. Our stray acquaintance continued to revert to the 'old days'—ten years ago! "She was a hummer, I tell you. This burg was the speediest town on earth. It makes me feel like an old man who knows he has lost his opportunities." Let not the moralist assume that it was the opportunity to be "speedy" that he regretted, it was the chance to become rich on which he chiefly lingered. But we refused to linger mentally with him. This 'might have been' bewailing the former tinseled splendors of a wide-open boom mining camp did not win our sympathy—only a humorous appreciation. The Dawson of today is as much better than the inferno of a decade ago as sanity is better than folly, as sobriety is better than debauchery. The quiet neat town of orderly people, under an incorruptible police, and a competent administration, is not for one moment to be compared with the rabble of excited adventurers and degraded women who made the Arctic a hotter place than the tropics. Let the panegyrist of the past go, with his regrets for wasted opportunity and his memories of besotted fooleries; in his stead is the intelligent engineer, the quiet man of business, and the orderly conduct of a civilized community.

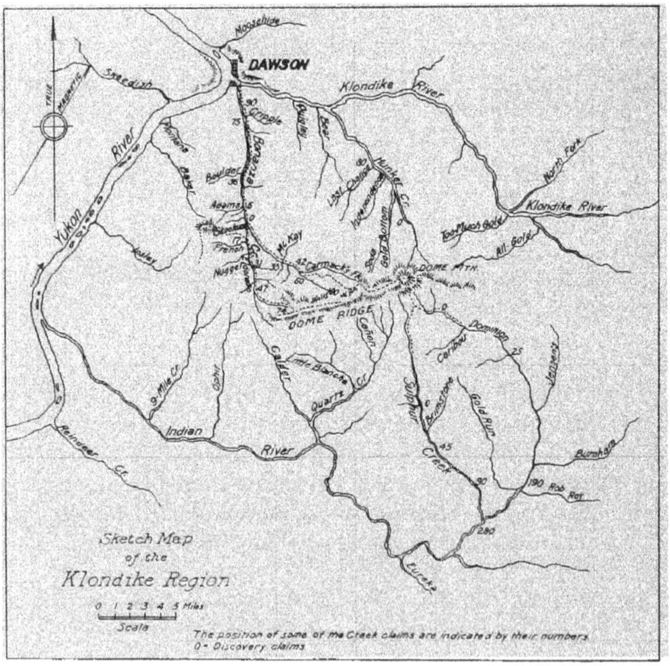

CHAPTER XVIII.

THE GOLD OF THE KLONDIKE.

When George Carmack found gold on Bonanza creek on August 17, 1896, he started a stampede of world wide interest. The rush to the Klondike, due to this discovery and to others that ensued, was the greatest wave of adventure since the days of 1849 and 1851, when California and Australia lured the gold-seeker to the conquest of the wilderness.

Two men figure in the story: Carmack and Henderson. Carmack was a fisherman, with an Indian squaw; he had a sort of trading post on the Yukon 20 miles above the Crossing. Bob Henderson was a nondescript prospector living at Sixty-Mile; he and three others found gold on a tributary of Hunker creek called Gold Bottom, in the summer of 1896. Henderson is believed to have been commissioned by Harper and Ladue, who kept a store at Sixty-Mile, now known as Ogilvie, to create local excitement by going through the motions of finding pay-gravel. In the history of the North this motive is frequent; when business became dull the owner of a store or trading post would send out prospectors to start an excitement for the stimulation of local trade. However that may be, Henderson went prospecting on the streams tributary to the Klondike river, and found gold on a branch of Hunker creek, which he reached by going up Indian river, on the south side of the water-shed, as is shown by the accompanying map.

At this time Carmack was fishing for salmon at the mouth of the Klondike, which enters the Yukon where Dawson now stands. Bonanza creek joins the Klondike two miles above the junction of the Klondike and the Yukon. Carmack, being short of fresh meat, went with his Indian companions, Skookum

Jim, Indian Pete, and Tagish Charlie, to Bonanza creek on a hunting expedition. Bonanza creek at that time was known to Indians and prospectors as a likely place for moose. Carmack and his friends knew that four white men were working on Gold Bottom, so they decided to cross the divide and pay them a visit, probably selling them some of the fresh meat they had obtained by hunting. Henderson and his partners were not getting much gold and Carmack soon returned to his camp. Having seen the mining done by the four men on Gold Bottom, he got the idea of doing a little prospecting himself on his return down Bonanza creek, and he actually found gold on the rim of the bedrock projecting above the creek. This rich spot proved later to be only a patch 20 feet square. He did not test the creek-bottom, for he did not know how, but he did make a location, and he knew enough to go at once down the Yukon to Forty-Mile, which was the nearest recording office. He recorded his discovery claim and also the claims located by the three Indians. He exhibited the gold he had found on the rim-rock, but the miners at Forty-Mile were sceptical as to the genuineness of the discovery. They had been fooled too often by discoveries concocted in the interest of store-keepers. But it happened that one wise head compared Carmack's gold with that of Forty-Mile, and he noticed a difference in the quality, for the Bonanza creek gold contains more silver. On this evidence it was decided that Carmack's story must be true.

A quiet 'rush' began. David Mackay, Daniel McGillivray, and Harry Waugh were the first to start. They staked No. 14 Below Discovery on Bonanza, where they themselves found gold on a shallow bar of gravel, known later as Poverty Bar, an intermediate or secondary bench deposit. Eventually each of these three men made a fortune. Andrew Hunker came up the river later in the fall, and staked a claim on No. 24 Below, and then went over the divide from Bonanza creek to Gold Bottom; from there he proceeded with his partner Johnson up what is now called Hunker creek, panning the gravel as he advanced. His idea was to go over the divide in order to reach the headwaters of a creek entering the Klondike some 25 miles above the mouth of Hunker, and even then known by the

CARMACK'S STAKES.
The posts that marked the Discovery on Bonanza Creek.

mythical name of All Gold. In one of his pannings he happened to find a reef of high bedrock exposed in the creek-bottom at the place now known as Discovery on Hunker, and from this spot he obtained over $25 in coarse gold. He and Johnson then staked three claims, namely, Discovery, No. 1 Above, and No. 1 Below. They went back, as they had come, by way of Gold Bottom, and told Henderson and his partners what they had found. Thereupon, a number of men, including the original four from Gold Bottom, staked rich claims on Hunker creek.

The excitement caused by Carmack's and Hunker's discoveries had spread throughout the upper Yukon; the few settlers at Circle City and Forty-Mile all came up the river and staked Eldorado, Hunker, Bonanza, and their tributary streams. Up to this time all the pay had been found on the upper reaches of the gulches and those who staked on the lower portion of Bonanza and Hunker were considered crazy. It was thought impossible that the run of gold could extend so far.

During the winter following these events, that is, the winter of 1896-'97, nearly everybody who had a claim on the creeks went to work and soon demonstrated the extraordinary richness of the ground. It was not until the summer of '97 that the high-level benches were investigated. In July Burk and Hensley made a discovery on the bench opposite No. 5 on Eldorado. One of the first men to start the bench diggings was Bill Gates, commonly known as Swiftwater Bill, because he was fond of talking about the kind of boat best fitted to go through swift water. He paid a man to prospect a high bench on the left 'limit' of 13 Eldorado, but unfortunately this man struck a spot where there was no gold to amount to anything. Skiff Mitchell was the next man to prospect on the high bench. In July he sent Lancaster and Stimson to do some work on Gold hill, opposite his own claim of No. 1 Eldorado. Lancaster and Stimson's two claims of 100 feet square each proved eventually to be marvelously rich, but they did not know it at the time they did their first work, in the fall of 1897. Probably $300,000 was taken out of these two small claims. There was a record of $2700 for one day's labor by two men, one trundling a wheel-

OLD GOLD CREEK.
One of the streams tributary to the Klondike.

barrow while the other manipulated the rocker. Lancaster had a letter of introduction to Mitchell and the hint to go up the hill was given in goodwill. At first, as stated, Lancaster and Stimson were not successful; but an immensely rich strike by Nels Petersen and N. Kresge on the other side of Gold hill, near the mouth of Skookum gulch, gave a fierce impetus to prospecting along the upper slopes of Bonanza creek.

Petersen and Kresge struck it on September 5, 1897. On the day of discovery they picked up $29 in coarse gold, and in 10 days they obtained $6375 by the use of a rocker from a claim 100 feet square. On the second afternoon they are said to have gathered $1100 in nuggets lying just under the moss. They extracted $12,000 from this claim and then sold it, in November following, for $40,000. Mr. Kresge, whom I met at Fairbanks, showed me a nugget containing $41 in gold mixed with double the proportion of quartz, the gold veining the quartz beautifully. The gold taken out by this prospector and his partner was found chiefly in the crevices of the bedrock; coarse gold was found as much as 8 feet down between the slabs of schist. Later workers penetrated 16 feet into the bedrock before it ceased to be profitable to dig deeper.

The success of Kresge and Petersen led Lancaster and Stimson to test the upper portions of their claim, where they also found rich pay. Shortly afterward William Dietering (or Caribou Bill, as he was generally known) located a bench claim on the other side of Skookum gulch, but he did not find much. He became short of grub and sold a half interest in his claim for a few sacks of flour and some bacon. His partner (Small) neither worked nor put a man to work, so they agreed to sell for $3000 to Chase and Campbell. These two prospected deeper into the bedrock, and at the first 'thaw' they took out $200. They sold the claim for $12,000.

Caribou Bill was an experienced miner from British Columbia and he had a good idea of the manner in which alluvial deposits are disposed. As soon as he had made a little money he grubstaked men to test the hillslopes for bench-gravels; thus he found gold on French hill, on Oro Grande, and on Little Skookum. Two other old hands from the Caribou, named Morrow and McCrimmon, sank a 100-foot shaft on the top of Gold hill and found rich gravel. This was in June 1898. Then the stampede for the hills became general and by the end of the summer of 1898 all the high gravel deposits were covered with locations and work had begun. Thus the wonderful terrace of white gravel flanking the slopes of Bonanza creek, down to the Klondike, became recognized under the name of the White Channel. A year later similar benches were found on Hunker creek.

And gold was found in astonishing quantity. On a day of June 1899 no less than 29 pack-horses came into Dawson from 'the creeks,' bringing gold belonging to one man, Alec Macdonald, a Nova Scotian, supposed at one time to be worth $7,000,000. But he lost most of it before he died—last winter.

Those who came to Dawson with the 'rush' at the end of 1897 and in the spring of the following year were too late. All the best ground had already been staked. They had to buy claims or work for wages. During the winter of 1898-'99 there were fully 10,000 idle and destitute men at Dawson.

Many stories are told of those days. Some of these are humorous. Thus, there was a Swede named Charles Anderson. He was an ignorant man living at Forty-Mile, where he had saved $600 from wages earned as a pick-and-shovel miner. Being drunk one day he was persuaded by two old prospectors to buy a claim, No. 29 Eldorado, for the aforesaid $600. The morning after the Swede found himself with a headache, less $600, and he begged for his money to be given back. He was frantic. But to no avail. The two perpetrators went prospecting on Quartz creek, while Anderson decided that the only thing for him to do was to test his claim. He said: "Ay tank Ay go to work." He traversed the 80 miles from Forty-Mile to his claim and then went 18 feet farther, to bedrock, where he found a fortune! Just at the psychological moment there came the two wise men from Quartz creek on their way to the little settlement at Dawson. They passed the Swede and saw that he was panning, so they stopped to ask him, in jest, if he had found anything. He answered: "Ay tank Ay got some gold here," and showed them a pan with $1400 in it. This claim yielded about $1,250,000; the purchase for $600 was a good one.

As a reward for his part in making the discovery of gold, with Carmack, the Indian known as Tagish Charlie was made a citizen by the Canadian government. This gave him the right to buy a drink, all other Indians being prohibited from doing so, for it is illegal to sell liquor to the natives. Tagish Charlie has always been proud to assert his unique privilege and loves to order drinks for anyone that happens along. In consequence, he spends much of his valuable time in the 'skookum house,' otherwise called a jail. Skookum Jim and Tagish Charlie were the brothers of Carmack's squaw. After living with this Indian woman for many years, Carmack abandoned her and his half-breed children. For this disloyalty he was disliked by the 'old timers.'

In the winter of 1902-03 on No. 28 Above on Bonanza, in a piece of ground only 40 by 60 feet, four 'lay-men' took out $208,000 on a 50% lay. The gravel averaged $10 per bucket (equal to 6 pans). Thus these four men got $104,000 in one winter. Three of them worked in the shaft and one above-ground, at the windlass. As the time was four months and the total cost of labor about $5000, the net profit was close to $100,000. By their agreement with the owner they could hire no more men, only four workers being allowed. These fellows went to the Tanana diggings and lost what they had made so easily on Bonanza.

The 'ladies of adventure' made lots of money at Dawson, some of them 'cleaned up' from $25,000 to $100,000. They got an interest in claims with the diggers and even married them. Others, returning to 'the States' bought land at Seattle and doubled their stake. Gambling, however, ruined many of the prostitutes, as it did their consorts; and with the dissipated of both sexes the money was lost usually when they were drunk. These facts are not romantic, but they throw a side light on the real life of a 'boom camp.'

Dick Lowe and his fraction contributed to the romance of the Yukon. When Ogilvie was running a base-line for the Dominion Government, his party of surveyors included Dick

Lowe. One day when Lowe was carrying the chain for Ogilvie the latter found that a claim-holder had staked more than the 500 feet allowed by law, and there was a fraction 86 feet 4 inches long that had thereby become subject to location. Ogilvie told Lowe to stake the 86 feet, but Lowe demurred at first, not wishing to lose his 'rights.' For no man could locate more than one claim in any one of the two mining districts, the Klondike and Dominion. However, Lowe did stake the fraction. He cleaned up $46,000 from the bedrock in 8 hours. He paid the Government tax on $346,000, and it is probable that this small claim yielded $500,000. It has been said that Dick Lowe took $750,000 out of his 86-foot fraction, but the smaller figure

TRANSPORT OF SUPPLIES.

given is more likely to be correct. The mine was run in a careless way. All sorts of people made clean-ups and gave away nuggets. One pan yielded $900. The ground was so rich that in order to prevent trespass a wire was swung along the boundary line between the Dick Lowe fraction and No. 2 Below; the ownership of a nugget lying on the line was determined by a plumb-bob sliding on the wire, the nugget going to the side on which the larger part of it lay. Nearly as much gold as was saved is supposed to have been stolen, for Lowe was drunk most of the time. He went to Fairbanks in 1905; but being unsuccessful, he soon became poor. He returned to San Francisco, where he died in 1907 at the home of Mrs. James Hall, who cared for him during his last days. Dick Lowe is said on one occasion to have thrown as much as $10,000 on the bar and invited all the riff-raff of prostitution to drink with him, but he never helped a friend in need. "He was just naturally *cultus.*" An expressive epitaph and an inexpressibly pathetic sequel to the winning of much gold.

In the spring of '98 there were but few things that could be bought for less than 50 cents. Eggs were worth $18 per dozen; potatoes, $1 per pound; beef, $1 per pound on the hoof. Ten head of steers, brought down the river from White Horse, sold for $11,000. The first chickens (weighing 4 pounds apiece) fetched $10 each. Two crates of cats sold at $25 per cat, although they had been collected at Vancouver by boys who received 15 to 20 cents for them. Labor cost $1.50 per hour and men earned $15 per day. But a pint of champagne fetched $15 also. At the Forks the diggers paid $1.50 for each dance with the 'ladies of adventure'—a dance that lasted two or three minutes. For this $1.50 the man got a drink and the girl had one also, the latter receiving a commission of 50 per cent of the payment. Men would spend $100 in a night in mere dancing, without ordering 'wine' or gambling. Beer sold for $7.50 per pint, and many were the fools that paid for it. Some of the brightest fellows lost their senses and after a hard day's work would plunge into silly carousals that cost hundreds, even thousands, of dollars. It was the old story of money easily made and easily spent; only a few men of character withstood the temptations of the gambler and the prostitute, retained the gold won by hard work, and returned to their homes, the stronger and the better for the experience. To them the Klondike rush was a truly romantic episode.

CHAPTER XIX.

THE DIGGINGS.

'Klondike' is a corruption of *Thron-duick*, meaning plenty of fish, for here the Indians had a fishing camp long before the gold was found. In 1887 Ogilvie reported: "The Indians catch great numbers of salmon here. A miner has prospected up this river for an estimated distance of forty miles. I did not see him." In 1896 Joseph Ladue built a shanty and started a store at the mouth of the Thron-duick or Klondike river, intending to use it as a branch of his trading post at Sixty-Mile. When the discoveries of gold were made in the summer of that year, the land on which he had squatted became valuable for town-lots. He made a fortune, of course. On July 14, 1897, the *Excelsior* reached San Francisco with the news of fabulous finds in the North. She brought half a million dollars in gold and was the first of the treasure-ships that entered the Golden Gate like Spanish gallions of the olden days. Whatever doubt there may have been concerning the truth of the stories brought by her passengers, was removed when three days later the *Portland* reached Seattle with more than $1,000,000 of gold on board. That was enough. The news was flashed around the world and from every quarter of the globe eager men rushed toward the Klondike in the hope of winning a fortune.

In the summer of 1908 there were about 6000 people in the district about Dawson, one-third of these being residents of the town. The mining activity of the Klondike and its tributary creeks had undergone concentration through the consolidation of claims acquired by the Yukon Gold Company. On the south side of the divide, that is, on Dominion, Sulphur, Quartz, and adjacent creeks, the ordinary small operations were being continued, but on the north side, along Bonanza, Eldorado, and Hunker creeks, the big company controlled by the Guggenheims had secured so much property as to diminish individual activity to insignificant proportions. In the meantime the Yukon Gold Company was carrying out important undertakings requiring lots of capital and the skill of the best engineers. Much was said about the Guggenheims and their representatives and many of the 'old timers' sneered at their free spending of money and the new ideas introduced into Northern mining. With such an attitude I have no sympathy, for, while holding no brief for the clever Hebrews who have won so dominant a position in the American mining industry, I know the engineers in their employ to be professional men of the highest standing and of ample experience.

The Yukon Gold Company illustrates many interesting features in mine operation and mine finance. After individuals and small syndicates had hastily garnered the gold in the richest claims and had proved the existence of gold-bearing deposits in the creeks and on the hillsides above them, an enterprising Englishman, named A. N. C. Treadgold, succeeded, by persistence and obstinate good sense, in obtaining options on enough ground to form the nucleus of a big mining enterprise on the Klondike watershed. Some of this ground he secured during the boom days, but, seeing the extravagance of the mining methods, he advised his English clients not to exploit their mines, but to await the reduction of cost and the improvement of method that always succeeds the first year or two of prolific production. When his English backers lost patience, he enlisted the financial participation of the Guggenheims, themselves aided in this venture by J. Pierpont Morgan, and thus the Yukon Gold Co. was organized in 1906. More claims were bought outright, options were obtained on others, examinations and surveys were quickly made, and a large scheme of operation outlined. By the end of 1907 not less than $8,220,000 had been spent and since then fully $3,000,000 more. In April 1908 the enterprise was discredited by a fiasco, 700,000 shares (out of the 3,500,000 shares at $5 par value) being offered to the public through the medium of flamboyant and

IN THE MINERS' BOARDING-HOUSE.

meretricious advertising on the part of an irresponsible stock-jobber named Thomas W. Lawson, of Boston. The issue was a fizzle, only about 300,000 shares were placed, at prices ranging about $6, and the quotation dropped quickly to $3.50. This flash-in-the-pan of speculation and attempt to use a legitimate enterprise for improper purposes hurt the name of the Yukon Gold Co., injured the reputation of the Guggenheims, and brought discredit on an undertaking of great importance to the future development of the Yukon Territory. It is my sincere hope that the vagaries of a queer kind of financial legerdemain will not imperil the success of an engineering work of the greatest utility and that it may be carried to completion under the direction of O. B. Perry, the consulting engineer, and Chester A. Thomas, the resident manager.

In this book there is no place for technical details, so I shall give only a general description of the work as I saw it. Anyone wanting technical information on the subject will find it elsewhere.* In order to find room for the 'tailing' or waste resulting from the mining of the deposits on the hillside it is necessary to own the valley below. Thus, before commencing to exploit the bench gravels it is imperative that dumping facilities be obtained by acquiring the claims along the creeks; having acquired these, it is necessary to work out the gold-bearing gravels contained in the latter before covering them with the débris from above. Hence dredging and other methods of working the creek gravel were started at once, with a view to extracting the gold and preparing for the washing of the bench gravel when sufficient water under pressure became available. To do this a conduit had to be built bringing water a distance of 70 miles and delivering it by pipe under a head of 350 feet. While the ditch was being built, an electric power-plant was erected and the energy transmitted to the dredges and other machinery by means of a line 35 miles long. In July 1908 I found seven dredges at work, a large reservoir had been finished, the power-plant was in service, the ditch

**Mining and Scientific Press.* August 29, September 12, and December 26, 1908. January 9, 16, and 23, 1909.

A MEANDERING STREAM IN GOLD-BEARING ALLUVIUM.

was approaching completion, and operations of a wide-spread and highly technical character were being conducted systematically but with feverish haste, both by day and by night.

On July 14 I had a first glimpse of the celebrated diggings. It was a fine sunny day, with the thermometer at 85°F. in the shade. We drove along the road built by the Territorial government, over the Ogilvie bridge, which spans the Klondike and is named after William Ogilvie, the first Commissioner or Governor of the Territory. Near the bridge the Williams or Bonanza Basin dredge was digging the bed of the Klondike; just above the point where Bonanza creek joins the Klondike, we saw the No. 1 and No. 2 dredges of the Yukon Gold Co., one digging and the other idle pending a clean-up. The famous valley belied its reputation. All was strangely quiet. The former workings are largely obliterated by a growth of brush and the destruction of buildings erected during the busy days. In a country where lumber and fire-wood are expensive, unused buildings are pulled down without delay for use as fuel. The 'creek' has the appearance of a mining camp of 50 years ago instead of one that was at the zenith of its glory barely 10 years past. Even claims worked two years ago look as if they had not been touched for twenty, for the winter snows and the spring vegetation quickly heal the scars made by man. It is difficult to realize that this quiet valley, between rounded green hillsides, even though marked by open workings, was lately the scene of fierce activity, intense hurry, and an astonishing production of gold. In those days the work of mining was done by digging holes in the frozen gravel, using wood fires to thaw the ice and soften the ground, so that it could be excavated with pick and shovel. If you had gone up the narrow valleys of Bonanza and Hunker during the long twilight of the arctic winter ten years ago you would have seen a picture worthy of Gustave Doré. There was no noise, for there was no machinery; there were no whistles to announce the noon hour or the evening rest; there was no drilling in hard rock nor cheerful hammering. A weird silence brooded over the waste of snow. The gloom was thickened by a pall of smoke escaping from holes in the ground, whence an occas-

ON DOME CREEK. YUKON.

ional figure issued. Not many men were visible, for they were below in the rabbit warren of their diggings. At the top of a shaft, here and there, a weary gnome might be espied turning a windlass and emptying buckets loaded with dirt that came from a small pit beneath. The flare of red fires parting the twilight marked the beginning of the work of shaft-sinking. The snow, the moss, and the fog muffled every foot-fall, deadened every sound. It looked like hell—but it was freezing.

In the foregoing account I have used several technical terms, the explanation of which will throw a light on the local conditions. At the time of the first discoveries a placer claim was 500 feet long measured in the direction of the creek, for the full width of the creek. These dimensions were subsequently changed so often as to intimidate speculators. The prospector who made the first find of gold was entitled to an extra claim, as a reward. This was the Discovery claim. From it all others took their station, thus the third claim above the Discovery claim was called No. 3 Above; the name of the creek being added, it became No. 3 Above on Bonanza. If below the Discovery, then it was No. 3 Below. The sides of the valley were termed the 'limits' of the creek, so that "the right limit" meant the right bank going down-stream. A royalty of 10 per cent "on the gold mined" was collected by the Government of Canada. No miner could receive "a grant of more than one mining claim in a mining district," but he could also stake one hill claim. This was done to prevent a few men pegging the whole country; and yet the law was evaded, so that the regulations were subsequently modified. I quote them as enforced during 1898.

In addition to terms connected with the ownership of mining claims, I have been compelled to use words strange to an untechnical reader; as I wish to hold his interest in a further description of the methods of mining, I shall furnish him with the necessary glossary. It is a courtesy that the author owes to those who mentally travel with him.

Each mining region has its own local terms, originating from the interplay of peculiar men and peculiar conditions. Some of these terms are expressive; indeed, they may be so expressive as to become a necessary part of a telling description. Other terms by their wide applicability become serviceable in regions beyond the place of their birth and pass into the linguistic heritage of the race. Others again are merely the vulgarisms of the moment or the provincialisms of uneducated men, and the sooner they are thrown over the scrap-heap the better.

The arctic moss that carpets the northern wild is called 'tundra'; this is a Russian word and was borrowed from Siberia. The dirty blanket of frozen mould that covers the face of the North is called 'muck.' Under a few inches of dull green moss there is a thickness, varying according to locality and exposure, of ice, in which are embedded fragments of roots, moss, mould, and rock débris. In a temperate climate this would mean a layer of soil; in the North, it means a much greater thickness of black ice, which thaws to a liquid mud. Fully 60% of the 'muck' is water, the remainder is mostly organic material light enough to float. It is present everywhere; and as it is ubiquitous in the topography so it is also omnipresent in speech. There is no synonym to replace 'muck'; mud will not serve, for it is not mud; 'mud' is moistened earth; 'loam,' 'soil,' 'mould,' and the like do not express the frozen condition. It is true, 'muck' signifies nothing to those who have not been in the North, but to a 'sour-dough' or old timer, it has a world of meaning, for it is the one great natural obstacle against which he has fought time and again.

The gold-bearing sediment forms part of the débris deposited in the former bed of a stream. The gravel is called 'wash'; the rock on which it lies is called 'bedrock'; and when it rises to the edge of the creek-bottom, it is known as the 'rim' or 'rim-rock.' The richest portion of the deposit is usually at the base of the gravel, on bedrock; and as it lies lengthwise with the course of the creek, in places being as wide as the valley, while in others restricted to a riband, it is called the 'pay-streak.' It would be better to call it a 'channel.' In Australia, this central portion of the gold-bearing alluvium is labeled the 'gutter.' The particles of gold are called 'colors,' if small; and 'nuggets,' if large. To ascertain how much gold the 'dirt' carries, the miner washes ten or twenty pounds of it in his 'pan'; this is a sheet-iron dish, nothing less than a frying-pan without a handle. By use and experience, it is now made of a shape best adapted to the rough process of concentration performed by the prospector when he gives it a shaking motion, aiding the water in the removal of all the lighter particles of rock and stone wherein the gold lies imbedded, until finally a string of yellow particles remains on the bottom of the pan.

The next device is the 'sluice-box,' a board trough into which the gravel is shoveled while a stream of water is admitted so as to wash away everything except the gold. This separation between the valuable and the valueless constituents is facilitated by cleats or 'riffles' nailed across the sluice-box to arrest the gold when it sinks to the bottom. A sluice cut in the bedrock is called a 'ground-sluice.' Water under pressure is used to move the gravel; in rudimentary practice this water is conveyed in a canvas hose and when the scale of operations increases the same service is performed by iron or steel pipe. The nozzle becomes a 'monitor' and the operation of driving the gravel with a powerful jet is hydraulic mining or 'hydraulicking.'

The following table of alluvial measures will be useful to those who are interested in Alaskan practice:

1 pan holds	25 lb. of gravel
6 pans	1 cubic foot
15 pans	1 wheelbarrow
10 wheelbarrows	1 cubic yard
135 pans	1 cubic yard
4 wheelbarrows	1 bucket

These do not agree exactly. A full pan will hold from 20 to 25 lb., and it requires from 125 to 135 pans to make a cubic yard. A cubic yard is usually estimated to weigh 3000 lb., or 1½ tons. If a pan holds 20 lb. and 150 pans equal a yard, then a cubic yard weighs 3000 pounds. A loaded wheelbarrow will hold one tenth of a cubic yard; this is the ratio recognized at Fairbanks and at Nome.

CHAPTER XX.

DEVELOPMENT OF MINING METHODS.

Gravel mining is a simple process; the simplicity of empirical deduction. It is the growth of experience in overcoming natural obstacles. I shall endeavor to describe methods that have enabled man to extract an astonishing amount of gold. The creeks at Dawson have yielded $125,000,000 in ten years, the alluvial flats of Fairbanks have given the world not less than $32,000,000 in five years, and the golden beaches of Nome have contributed fully $22,000,000 in eight years.

This gold has come for the most part from deposits of gravel lying in or beneath the beds of existing streams, meandering within the limits of shallow valleys. Here is a typical example: A small valley overlooked by rounded hillslopes is traversed by a stream the present bed of which is only a few yards wide as compared to the flat, half a mile wide, over which it wanders. Bare ground, in the form of gravel, is visible only on the edge and in the bed of the stream; the remainder of the valley is covered with moss, out of which arise clumps of spruce, some a foot in diameter. On the hillsides the forest grows scantier, and on the summits the ridges are silhouetted in sweeping lines unbroken by any trees. A few specks of gold are found in the gravel, and there are rare spots where the rim shows coarse 'colors.' The bedrock is probably a soft schist, for that is the formation exposed by landslips; elsewhere it is covered by moss, by 'muck,' or by gravel. The prospector cannot sink a pit or shaft in the bed of the stream because the water will drown his workings. No pumps are available, nor is it feasible to divert the creek by means of a dam, because that would bank the water on another man's claim; moreover, the gravel is so porous that the water would penetrate into any workings sunk in the bed of the stream. Thereupon the prospector turns to one side and digs into the valley-bottom at a safe distance from the present stream-bed. He finds that under the moss the ground is frozen solid; it is impossible to sink a shaft with a pick, and expensive to do so with explosives. Yet he reasons that the indications point to the existence of gold on the bed of the valley to one side of the present stream, along a course that it formerly followed. What is he to do?

Such was the problem confronting the pioneers in the Yukon and Alaska ten years ago.

Machinery was lacking. The pick and shovel were the only tools available. Wood was handy. What more natural than to overcome ice with fire, to soften the frozen ground by artificial thawing? This was the pioneer's method. He laid his bundle of sticks and made a fire that melted the adjacent ice. In this way he sank a small shaft to bedrock. The work of sinking was done in winter, when surface-water did not impede. After the wood had burned so as to soften the ground, he broke the latter with his pick and hoisted it to the surface in a bucket with a windlass. Then he piled the gravel near the shaft's mouth in a heap, which re-froze during the winter and thawed naturally in the spring. As it thawed, he shoveled it into a sluice-box and washed it by the help of any water available. His whole equipment consisted of a pan, pick, shovel, a bucket made out of a whiskey-barrel or a hide, fire-wood, a hemp rope, two or three sluice-boxes each 10 or 12 feet long, and muscle, and more muscle, and persistence. It is wonderful what some of the pioneers accomplished. Thus Sam Samson and a partner, in the winter of 1901-02, sank a shaft 115 feet on the Cyrus Noble claim, near Nome. The shaft was only 2 by 4 feet. It was on the 'tundra.' There was no forest to yield good firewood, but Samson found scrub willows nearby and he burned them. He would fill a gunny-sack with willow twigs, dry them in the oven of his stove, and place them in the bottom of the shaft, under cover of a wash-tub, to retain the heat. A fire in the morning and another in the evening sufficed to soften the frozen gravel. He had to conserve the air in his shaft as best he could. He worked in his undershirt, perspiring while the air at the surface was below zero. After sinking 80 feet without the safeguard of timber, he lined the shaft to the surface with inch boards. This was an exceptional case, but it illustrates that grit can overcome gravel, even when frozen.

The gold is found concentrated upon the bedrock. This concentration is more complete in Alaska and the Yukon than

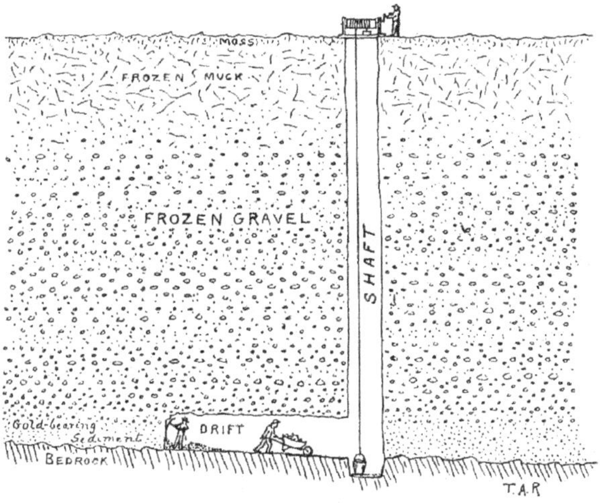

DRIFT-MINING IN FROZEN GROUND.

in other mining regions; it is due to the clean character of the 'wash,' that is, there is so little clay in the gravel that the descent of the heavy gold has not been hindered. It has fallen to the rock-bottom and lies there, sometimes so thick that the mass consists of more gold than dirt. In most cases the miner finds his 'pay' confined to the stuff that lies for a couple of feet above bedrock, and within the bedrock itself; for the gold has sunk into the crevices of the rock, penetrating sometimes three feet, if the schist be blocky and shattered. Therefore the operation of mining includes the removal of the bottom layer of sediment on the creek-bed and the top of the rock-bottom. From 2 to 6 feet of material is considered rich enough to be treated as 'pay' and is washed in the sluice-boxes.

While I speak of the rock-bottom, it must be explained that the bedrock is usually soft. It has undergone disintegration; it has been shattered by alternations of frost and thaw, in a by-gone time; it has been penetrated by water and some of its constituent minerals have been so dissolved as to leave it no longer hard and resisting, but docile as clay. The soft, almost 'mushy,' bedrock of the North is a great aid to the miner. He does not need explosives.

When the shaft reaches bedrock, it is an exciting moment. The miner scans the ground to find specks of gold; if the gravel be rich, he can see the gold readily. Then he hoists some of the soft bedrock and the fine sediment lying on its surface. By the use of a pan he washes this material and ascertains how rich it is. Often he sees a glittering string of yellow particles in his pan; sometimes a piece big enough to be called a 'nugget'; sometimes—nothing, only a little black sand. As an illustration of the extraordinary richness of some of these deposits, I cite the following: In August 1899 on No. 2 Above Discovery on Bonanza, adjoining the Dick Lowe fraction, George T. Coffey took two shovelfuls, that is, enough to fill a pan, and from it he washed 63 ounces of gold. This included three pieces worth over $100 each. It was possible to see the gold in the gravel when standing 20 feet away. Among those present on that occasion was Angelo Heilprin.

Ordinarily, 10 cents worth of gold, or 2½ grains, per pan, indicating a yield of $13.50 per cubic yard, say, one yard deep, was rich enough to yield a handsome profit to a man who sank a shaft 40 feet to bedrock.

If the shaft does not 'bottom' in pay, the prospector begins to explore laterally by digging a gallery, or 'drift,' following the surface of the bedrock. The shaft may be off the line of the maximum concentration—it has been sunk to the rim of the channel rather than the gutter—and a short drift will enable the miner to find better stuff. Whether he explore for richer pay or open out into a beautiful layer of golden

STEAM-POINTS IN PLACE UNDERGROUND.

sediment, he extends a drift from the bottom of his shaft and removes the ground by thawing with fire, as he did when sinking. The removal of gravel by hand-labor in this manner is called 'drifting,' as against methods in which water is the prime agent.

The thaw affects only a small patch of ground; it does not endanger the worker, who burrows patiently under a hard roof of frozen gravel. Bit by bit all the gold-bearing dirt within the boundaries of the claim is excavated and raised to the surface, to be washed in the sluice-boxes whenever water is available. This method of working frozen ground by thawing with wood fires was originated in the Forty-Mile and Circle districts before the discovery of the Klondike. Even where timber is cheap, it is more economical to exploit frozen ground in this way than to operate in thawed gravel.

In the days before Columbus discovered America the ancient miners of central Europe employed the method known as 'fire-setting.' A big wood fire was built close to the face of a level, and when the rock had become thoroughly heated it was customary to throw water on the hot surface, causing it to crack. When thus fractured, the ore was extracted by the further aid of hammers and wedges. According to Henry Louis, this method was in use in the Sala mines, in Sweden, as late as 1876, and in the Kongsberg mines, in Norway, it was employed up to 1884, when improvements in blasting caused the abandonment of the ancient practice. I have seen many a face of an old level in the Alps, on the border of France and Italy, that was beautifully concave by reason of the application of this method. On the other hand, in the copper region of Michigan the country is occasionally covered with a blanket of sand, wash, or gravel, which must be penetrated before the hard copper-bearing rock is reached. Sometimes the shaft breaks into a quicksand, making further sinking impossible in the ordinary way. Then artificial freezing is employed; the wet sand is frozen solid and kept in this condition long enough to allow the miners to make the necessary excavation and timber it securely. Thus man uses fire and frost, air and steam, wood and iron, in his subterranean operations, overcoming Nature in one place by the use of the very force she uses to resist him elsewhere.

Wood fires make smoke. The gases liberated are injurious to health. In the North, men soon learned to keep away from the shaft or drift until natural ventilation had purified the air. At best they had to work in a warm moist atmosphere, for they had to excavate the rock softened by thawing before it could freeze again. At the surface the air might be 20° below zero; in the mine the conditions simulated a Russian bath.

After the pioneers in the North had used wood fires for thawing during one winter season or more, a clever operator hit upon the idea of employing steam for the same purpose. Then the 'steam-point' was introduced. It happened thus: In

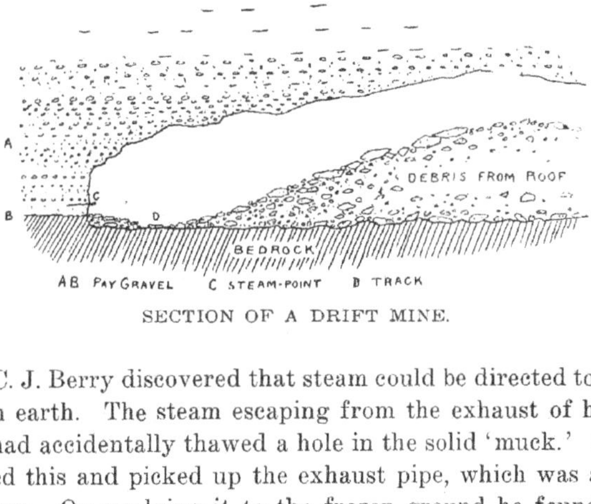

SECTION OF A DRIFT MINE.

1898 C. J. Berry discovered that steam could be directed to thaw frozen earth. The steam escaping from the exhaust of his engine had accidentally thawed a hole in the solid 'muck.' Berry noticed this and picked up the exhaust pipe, which was a rubber hose. On applying it to the frozen ground he found that it would thaw the muck so as to penetrate for the full length of the hose within a few minutes. This excited the men who happened to be watching the experiment. All of them at once began to devise a scheme for doing this work effectively. A rifle barrel was chosen, then a small hole was bored into one side so as to admit the steam. Thus the 'steam-point' was invented. In its rudimentary form the steam-point was a short length of iron pipe, pointed at one end, and attached to a length of rubber hose, through which steam traveled from a small boiler at the surface. The pointed end of the five or six feet of iron pipe was inserted into the frozen gravel and driven forward gently by taps from a hammer, as the ground was softened by the steam issuing from the orifice at the lower end. As finally developed, the steam-point became a specialized tool of great efficiency. A solid head was added to the end that is hammered and a protecting ring was welded to the forward end; the shank itself was made of pipe of the strongest kind, and armored rubber tubing replaced the ordinary hose. The length of the 'steam-point' ranges from 6 to 16 feet, the usual size being 8 feet. This is driven home so as to make a hole about 6 feet deep. The accompanying sketch will illustrate the details of construction. The hole in the steel head allows for the insertion of a bar, wherewith the point is turned so as to aid advance. As the operator hits the head with a hammer, he turns the point by means of a bar held in the other hand.

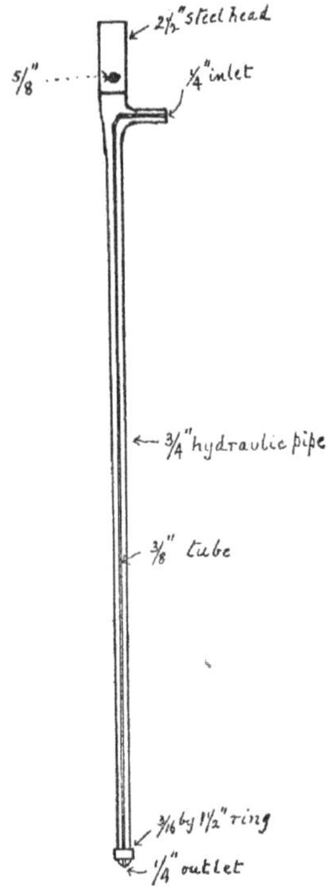

A STEAM-POINT.

As used two or three years ago, the cost of thawing was 25 to 30 cents per cubic yard. With longer points, longer 'sweating,' cheaper fuel, better system, the cost has been reduced one half. In a 20-ft. deposit, using 12 to 13-ft. points, it is possible to thaw 3½ to 5 cubic feet per point at each setting.

The efficiency of a point will vary according to the pressure of steam, the length of the tool itself, the distance between the points, the time allowed for 'sweating,' and the amount of moisture in the ground. An effort is made to fix the intervals between points so that their sphere of influence do not overlap.

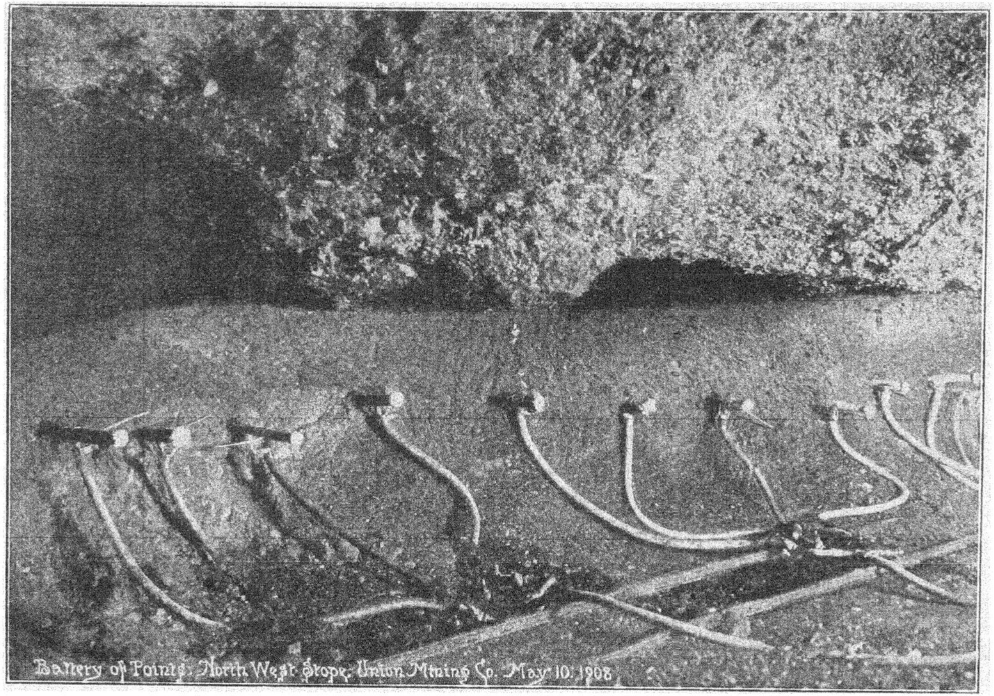

STEAM-POINTS IN ACTION UNDERGROUND.

By allowing the steam sufficient time to do its work, the area affected is increased. This is the 'sweating' stage. As the amount of humidity, in the form of ice, increases, more steam is required to overcome the latent cold.

When the steam-point was introduced, the extraction of gravel from the drift-mines was continued in summer, as well as winter, and the production of gold proceeded concurrently, as long as the weather at the surface permitted. The dump accumulated in winter would freeze before spring, necessitating the employment of steam-points before it could be moved. Moreover, the boiler erected for the purpose of thawing was also used for hoisting. Larger buckets and a bigger scale of operation became possible.

Thus, the frozen condition of the placers in Alaska and the Yukon, at first an insurmountable obstacle, proved, in the end, an aid to mining. To sink a shaft in the creek deposits of a warm climate means a persistent contest with water, for which pumps are necessary or a costly drainage system. The loose ground requires careful timbering. Some of the best portions of the channel may be unworkable because of an excessive influx of water. All this would have checkmated the diggers of the North in the early days of discovery. Pumps were 1500 to 2000 miles away, heavy timbers were scarcely to be obtained in most localities, a fight with water would have discouraged men unused to mining, as were most of those that rushed to Dawson, Fairbanks, and Nome.

The 'frost,' indeed, was the miner's friend. It enabled him to sink a shaft even in the bed of the creek; it permitted him to dispense with timbering; it allowed him to burrow with safety and to follow the layer of golden gravel with impunity under the ice-bound surface. Moreover, it obviated work on a large scale. One man could, and sometimes did, work alone, descending the shaft, filling the bucket, ascending to the surface, hoisting the load, and so forth. No machinery was needed save the simplest tools; no organization was required, beyond a willing partner; no capital, save muscle.

By the method of 'drifting' only the bottom layer of the gravel deposit was mined; any gold in the overburden re-

DREDGE NO. 6 OF THE YUKON GOLD COMPANY.

mained untouched. As all the gold is not concentrated on the bedrock, or in the five or six feet immediately overlying, this method entailed waste. It was not complete. Also there was danger of a collapse of the ground when warm air entered the mine in summer or by penetration of surface water into the workings. Thereupon the open-cut method was introduced. By this system the deposit was worked from the surface downward, the ground being removed in successive slices by 'scrapers' pulled by horses or by a steam-engine. The portion rich enough was taken in wheelbarrows or in a self-dumping bucket to the sluice-boxes, where the gold was extracted. Thus another method became successfully established.

Even this method was not applicable everywhere. In wet ground, in gravel partly thawed, and in the channels of live streams, neither 'drifting' nor 'open-cut' practice served the miner's purpose. He had to overcome a new difficulty. And he did it with the dredge, which had been already applied successfully in New Zealand and California.

A dredge consists of a bucket-elevator placed upon a barge. The barge is constructed at the bottom of a pit excavated, by the use of scrapers and horses, to a depth sufficiently below the expected water-level to ensure flotation and afford room for movement. Then the machinery is placed in position on the barge. As water is admitted, the dredge floats, and when it starts to work it digs its own way, filling the pit behind as it advances in the course of digging. The digging is effected by a chain of steel buckets, which excavate the gravel in front and deliver it to the washing apparatus at the rear of the scow or pontoon. After the gravel has been washed, it is discharged by an inclined traveling belt, which throws it sufficiently far behind the dredge as not to impede flotation or to permit of the same material being raised again by the buckets. As the ground is mined, the dredge is advanced forward and sideways by means of winches and ropes moored to posts on shore. The machinery is actuated by steam from boilers on the scow itself or by electric motors obtaining their energy from a plant at a distance. The hull is from 35 to 40 feet wide, 90 to 125 feet long, and 8 to 10 feet deep. The total weight is

700 to 800 tons. The buckets have a capacity varying on different dredges; 5 cubic feet is now deemed small; 7 cubic feet is common; 13 cubic feet is the maximum, as yet. If the buckets are of 7-foot capacity and they discharge at the rate of 23 per minute, the dredge when digging to a depth of 14 to 16 feet will handle from 90,000 to 115,000 cubic yards per month, allowing for unavoidable stoppages and repairs. The lips of the bucket are reinforced with manganese steel, to withstand the wear and tear. As they discharge at the upper end of the 'bucket-ladder,' the gravel falls either into a revolv

DREDGE ON BONANZA CREEK.

ing screen or upon a shaking table, with jets of water playing upon it so as to detach any clay and to facilitate the separation of the gold, which falls (through perforations in the screen or the table) on a series of sluices, provided with gold-saving devices, such as riffles, matting, and mercury.

Apart from economic conditions, which vary all over the world, the distinctive feature of dredging practice on the Yukon is the necessity for overcoming the frozen condition of the ground. This is a geologic frost as distinguished from the seasonal frost; the first is the sequel of a Glacial period, that is, a time when the summer thaw was unable to overtake the winter frost; the second is due merely to winter cold. The seasonal frost, as measured on ground that has been thawed in summer, reaches down 3 to 5 feet, but the geologic frost extends to a depth of 230 feet, if not more, wherever the ground is wet, as in the valley-bottoms. On the other hand, the warmth of the short summer, lasting 4 to 4½ months, will thaw the ground to a depth of 4 to 7 feet, according to local conditions, the chief of which is the nature of the surface-covering. In two seasons the frost in a gravel deposit may be conquered by the seasonal thaw to a depth of 10 to 22 feet. Even in the coldest winter, such as that of 1905, when the thermometer registered a minimum temperatude of –71° F., the frost did not overtake the summer thaw—on ground that had been 'stripped'—by three feet.

The Northland is covered by a blanket of moss and loam due to the decay of vegetation. This overburden is called 'muck' by the miners because when it melts it runs like thin mud, for it is composed of 25 to 40% organic matter and 60 to 75% ice. When either the summer sun or artificial heat strikes this black blanket, it disintegrates and is readily floated on the running stream. Being a mixture of organic matter and ice, it makes a perfect insulator against heat, and protects the underlying frozen ground from the warm air of summer. The thickness of this frozen muck varies from a few inches to 40 feet, the maximum being in gullies where it has accumulated by sliding from adjoining hillsides. Two feet is an average thickness. In summer, it melts; in winter, it freezes solid.

It is obvious that before the seasonal thaw can become effective, the blanket of moss must be removed. This is done naturally by freshets and by meandering streams; it has been done systematically both by ground-sluicing and also, much more rapidly, by hydraulicking. Where time is not an immediate factor and where bedrock is not more than 15 feet deep, a gravel deposit can be thawed to bedrock in two seasons by simply removing the cover of moss and loam so that the summer heat may get an opportunity to penetrate. But, on the other hand, where dredging operations cannot wait for such a slow method, or bedrock is deeper than 15 feet, or where old workings exist, it becomes necessary both to accelerate and to perfect the thawing process by the aid of artificial methods. This is done by the application of steam.

By introducing steam into the frozen ground, the ice is melted; the more water (that is, ice) there is in the gravel, the more steam will be consumed in converting water from its solid to its liquid state. Ice is a non-conductor; rock is relatively a good conductor, therefore, the less water the gravel contains, the more easily is it thawed. The stones retain the heat imparted to them so as to radiate it slowly into the surrounding mass. The method of thawing by steam was exemplified on claim 90 Below Discovery on Bonanza creek, just above the No. 6 dredge of the Yukon Gold Co. The plant consists of three boilers, fired with wood at $8 per cord delivered, the steam-gauge registering a pressure of 150 pounds. The main pipe-line is 3½ inches in diameter, and is carried within boxes packed with sawdust as an insulator. All the pipes are wrapped with 'asbestos' covering. The branch pipes approaching the ground to be thawed (at 300 to 400 feet from the boiler-plant) are 1½ inches, also insulated and boxed. At intervals of 8 feet, openings in the pipe connect through nipples with short lengths of hose. This hose is ⅝ inch diameter, and 17 feet long; it must be long enough to reach the heads of the 'steam-points' when they are being swung into position, and to allow latitude in twisting the 'points' The 'point' is a steel pipe ½ to ¾ inch, made in lengths of 14 to 20 feet, one end of which is hammered while the other end is being driven into the ground. The point itself, or advancing end, has an orifice 1/16 inch, and through this the steam enters the ground. By the time the steam reaches the place where it escapes into the gravel the pressure has sunk to 25 pounds per square inch.

In starting a point it is customary to take a 1¼-inch steel bar and drive it down with an 8-pound sledge-hammer until frozen ground is struck; then the steam-point itself is introduced, the bar having been withdrawn. The top or head of the point is pounded by one man with a 4-pound hammer, and at the same time a twist is imparted by a wrench held by another man. This is done at intervals as the ground is softened by thawing, the man in charge proceeding from one 'point' to another, in rotation. The rate of thaw is about two feet per hour. If the point refuses to sink, it is allowed to rest for a while, and then if no progress can be made in the usual way, it is inferred that an obstruction, such as a boulder, exists. The steam-pipe is pulled out of the hole with the aid of a lifting-jack and a solid steel bar is introduced; this is then hammered with a sledge so as to penetrate the obstruction or push it aside. If the incompetence of the steam-point is due to the clogging of the orifice through which the steam is emitted, this is followed by condensation at the head, as indicated by chilling of the pipe, and also by listening for the flow of steam through the hose. I noted that frozen ground was struck in most cases at six feet, this being the depth to which the seasonal thaw had progressed after the moss had been removed.

In this manner the ground is softened and made ready for dredging. The thawed ground will remain warm for a month, but it is not advisable to apply steam-points too far ahead of the dredge, lest the material should freeze again when the weather becomes wintry, nor to dig into the ground until the artificial heat has been fully expended.

Space does not permit, nor the occasion warrant, a detailed description of the process of dredging. Let me take you aboard the No. 6 dredge of the Yukon Gold Co. for a few minutes. It is doing well; crunching, groaning, and squeaking in the throes of laborious exertion, the machinery is digging into the gravel with relentless power and raising the gold-bearing dirt into the big revolving screen, where the pebbles make a great roar as they are carried a quarter turn before being ejected upon the rubber belt of the conveyor. This carries the boulders, pebbles, and roots to the pile of reject in the rear. You can climb to the extreme end of the 'stacker' or conveyor and watch the reject issuing from the interior of the dredge in a steady stream. Then go to the winch-room and note how a single man controls the huge mechanism by means of a few levers; there the vibration and straining of the dredge have a

THAWING WITH STEAM-POINTS AHEAD OF A DREDGE.

meaning, expressing the energy expended in digging the gravel of the creek and the power consumed in moving the gold-bearing material through various stages of treatment. The observer will be impressed at first with the tremendous strain incident to the operation, and then he will begin to share the confidence of the dredge-master who knows that his machine was designed to overcome all and any of the obstacles presented from moment to moment and from day to day.

We saw the clean-up, no less than $10,500 being collected as the result of 44 hours work, or at the rate of 85 cents per cubic yard, at a cost of 25 cents per yard. This was better than the average, but it indicated a profitable type of mining.

A HOME IN THE NORTH.

CHAPTER XXI.

ON BONANZA CREEK.

On June 21 at Dawson, which is in latitude 64° north, the sun sets at 11 p.m. and rises at 1 a.m. As seen from a height (such as the Moosehide Dome) the sun disappears below the horizon, but the rays do not become extinguished, so that direct sunlight is appreciable even at midnight. At the close of December the sunshine lasts only from 10:30 in the morning to 2:30 in the afternoon and the light is feebly actinic. Artificial illumination is used during 20 hours, yet, owing to reflection from the snow, no inky darkness supervenes, the black nights being those that precede the winter before the snow has covered the ground. During July it was never dark enough at any time to reveal the stars, but in the first week of August I saw them again at Fort Gibbon, like the face of a friend. To emphasize the quality of the light up to a late hour in midsummer, I can state that on July 19 I heard a surveyor complain that at 11:30 on the previous night he had been unable to see the cross-hairs in his telescope—half an hour before midnight! The long day enables work to proceed throughout the 24 hours and it turns the short season of 4 to 4½ months into a period of intense activity. The air and the light both favor continued exertion; men lose the habit of sleep, and, like bears, postpone arrears of slumber until the hibernating season. To those unaccustomed to these conditions, the difficulty of excluding the brilliant daylight from a bedroom, bunk, or tent at 10 or 11 o'clock at night prevents restful sleep. Dawson is quiet until 10 or 11 in the morning; the best time for business engagements is after the conventional dinner-hour, in the evening. Those who have lived in the North for many years tell me, as one might expect, that the habitual disregard of regular hours for sleep affects the nerves; but to the visitor, it is different; to be wide awake, mentally and physically, to feel the exhilaration of an air as of the morning of Time, and to receive the stimulus of an atmosphere charged with ozone is to realize the difference between man and a vegetable.

The fact that Dawson is no longer on the ragged edge of civilization was emphasized by the sight of two automobiles. One of them belonged to O. B. Perry, the consulting engineer of the Yukon Gold Company. With him I made several runs up the creeks over the corduroy roads. Never seemed the wilderness so vanquished as when we careered rapidly over ground impassable a few years ago to vehicular traffic of any kind. The machine was a Peerless three-cylinder roadster, specially designed in body and equipment for local conditions. It seated four persons. The body was made narrow, and stiffened by being built entirely of oak and aluminum. Gasoline sells for 90 cents per gallon, while hay costs $91 per ton, so that the motor consumes less money per mile than the horse. Time is worth more than either; to the engineer directing work in widely separated localities some mode of swift transport is essential, and the most rapid is the most economical.

Under the guidance of George T. Coffey, I visited the workings of the mines situated on the bench or terrace of gravel called the White Channel, which is 195 feet above the surface of Bonanza creek. Into these hillslopes the gold-seekers dug, making a warren of activity during 1898 and 1899. Men, singly and in partnership, drove tunnels into the gravel and extracted the gold-bearing portion by thawing with wood-fires and steam-points; then, using pick and shovel, they brought the 'dirt' to daylight in wheelbarrows that were discharged near the sluice-boxes. Feverish was their haste, for over many of them hung, like a sword of Damocles, the fear of insecurity of title; some desired to clean-up in a hurry and go home, others were possessed with the gluttony of work, or were spurred by a greed that knew no limit.

But when I saw Cheechako hill on a July evening in 1908, all was peaceful. No one was at work. Before me stretched an

IN THE EARLY DAYS. WASHING THE GOLDEN GRAVEL.

unassertive topography: rounded hills and gently sloping valleys, devoid of rocky outcrops, green with rank grass and berry bushes, the purple patches of wild flower merging into the blue of the far horizon. A straight line of weather-beaten flume, the big scar of a hydraulic mine, a few drab dumps in the gully, one or two neglected cabins, the winding road—only these suggested the irruption of man into the peaceful wild.

The "purple patches of wild flower" in the preceding paragraph refer to the pink fireweed or *angustifolium*, which grows wherever bush-fires have passed. This 'fireweed' is the characteristic flower of the Far North and makes brilliant splashes of color in the monotonous dark green of Arctic verdure. Yet, like the 'barrens' of the Labrador and the moors of Nova Scotia, the Alaskan wilderness is rich in bushes that yield edible berries. Raspberries and blueberries are plentiful, also red currant, strawberries, and salmon-berry (*rubus spectabilis*). But labor is so costly that each one must pick his own dessert. Canned raspberries will be supplied at road-houses and in camps when the neighboring hillsides are burdened with a better fruit, but it is cheaper to serve the canned variety that has been transported 2000 miles than to pay a boy to gather the berries. Similar improvidence is suggested by the fact that on the steamboats they serve canned salmon within hearing of the splash made by magnificent fish of the same variety in the river alongside.

In Skookum gulch, where the gravel had been removed and the bedrock exposed, we saw tusks of the mastodon, 12 to 15 feet long, lying white and bare; also horns of the buffalo and musk ox, washed from under the tundra in the intermediate transverse gullies. On enquiry, it was ascertained that in the gold-bearing gravel the miners found the tusks of the mammoth, as well as the skulls and jawbones of the mastodon and the musk ox. In the White Channel no such remains have been discovered, suggesting that the bench deposit is of Pliocene age, or older than the creek gravel enclosing the mammalian relics. The bones are frequently seen scattered on the bedrock of abandoned mining claims; teeth of the mammoth weighing 20 pounds apiece were on exhibition in Dawson, as well as

tusks 8½ to 9½ feet long and 6 inches in diameter. It is related that on Hunker creek when thawing the gravel the steam acting on this pre-historic cemetery caused an awful stench.

Stories of the mammoth, and mammoth stories, abound in Alaska. At least three authenticated cases are recorded of the finding of portions of the hair and the hide, with the bones, of the big beast, but no whole body has been found. The hope of such a find is based on the fact that a mammoth was found encased in the ice of the morass bordering the Lena river in Siberia and the carcass was extricated, skinned, and stuffed; it is now an exhibit in the Museum of Natural History at St. Petersburg and constitutes one of the great curiosities of the world. The mammoth flourished 25,000 years ago, and the body found had been in cold storage for centuries. The interior of Alaska was not covered by the great ice sheet of the Glacial period, the small precipitation preventing, and thus failed to provide means for preserving the remains of a time when man was beginning to assert himself in the scheme of creation.

Returning to Cheechako hill, and examining the foreground of the view, I saw traces of the frenzied men who dug eagerly for gold and gophered the hill with their tunnelings. Tools lay scattered in reckless abandonment: picks that must have cost from $10 to $20, that is, from half an ounce to an ounce of gold, saws that are worth even now $6 at Dawson, nails that were bought at $100 per keg; car-wheels, wheelbarrows, and other implements of mining—all these were rusting on the wet ground and suggested something of the extravagance of the 'early' days.

At that period Grand Forks, the little settlement at the junction of Eldorado and Bonanza creeks, was a pandemonium of drunken debauchery, where a successful miner would spend as much as $2000 in a single night, ordering 'drinks' (that is, pints of champagne) for two or three women at a time and telling the waiter to subtract the required amount of 'dust' from the small moose-leather sack, called a 'poke,' in which the gold was carried. The owners of mines passed the night in drinking and dancing, sleeping in the day-time, while 'laymen' on their claims were digging the money for these pro-

FLAT CREEK, A TRIBUTARY OF THE KLONDIKE.

tracted sprees. Miners were paid $15 per shift of 10 hours; thus a shift was worth the price of a pint of champagne, a relation suggestion that the crude political economy of the mining camp has an 'index number' of its own. Another unit was the cost of mining per yard: the output of a man engaged in drifting or open-cutting with pick and shovel averaged 5 cubic yards per shift, so that, including miscellaneous expenses, it was not profitable to exploit ground yielding less than $3 per cubic yard.

All the best ground on the rich creeks had been located before the 'rush' arrived. Late in 1896 and during the ensuing winter the miners of Circle City and Forty-Mile arrived and secured claims, so that by the end of 1897 fully 500 of the old settlers had planted their stakes. When the crowd arrived from the 'outside' in the spring of 1898, and in the summer following, they found themselves too late. Many of them went home immediately, disillusioned and disgusted. A few that had seen something of mining, among whom were experienced men from California and Australia, persisted in their search and found the bench gravels that had been overlooked by those who came first upon the ground.

Englishmen and Americans are prone to compare the methods of government under their respective flags, and while I was at Dawson I took pains to interview a few thoughtful Americans in the hope of eliciting their opinions concerning the local administration. In the main the verdict is favorable; at the present time the Yukon is efficiently and honestly governed, although when the gold excitement was at its height there was some corruption; that was inevitable. From the winter of 1898 to 1903 there was jobbery in the Administration building, and a law had to be passed forbidding government employees to hold claims. After that the officials used go-betweens, but carefully. If a fraction was found to exist between claims, by reason of careless staking, and if then this vacant ground was located and the locator went to the Recorder, he was likely to be told that the ground had already been located; thereupon someone was tipped to go to the creek and plant his stakes, unless a half-interest was given by the locator to some of the friends of the corrupt official. Blanket concessions also provoked resentment. In 1902 Treadgold applied for a concession of all claims reverting to the Crown in the entire Klondike district, undertaking to install a comprehensive water system in return, but, after the lease had been actually signed by the Minister of the Interior a public meeting

THE PROSPECTOR AND HIS ROCKER.

was called at Dawson and the indignation expressed led to the cancellation of the agreement. Nor did Treadgold sue the Government.

In '97 and '98 the impression prevailed among the diggers that the Government would not renew the leases for another year, it being expected that the claims would be reduced from 500 feet to 100 feet square, as was done in the Cariboo district.

This made them anxious to get out what gold they could before anything happened, for it was known that an Order in Council could modify the local regulations at short order. This fear led to reckless methods, which, coupled with the high cost of transport on supplies and equipment, made the cost of mining most extravagant. Freight was 25 cents per pound. Thus the claim known as No. 2 Above on Bonanza yielded $800,000 between 1898 and 1902, yet only $37\frac{1}{2}$ per cent in dividends was paid on a capital of $400,000. If worked 10 years later—in 1908—when freight was 1 cent per pound and ordinary labor $5 instead of $15 per shift, the profit would have been $600,000.

Of the claim-holders on the rich creeks fully two-thirds were American citizens from the old camps down the Yukon; moreover, after the rush there were 60,000 people in the district, and only two rich creeks to divide among them; thus, any sub-division of claims would have been popular among the disgruntled majority. When Ogilvie talked about the marvelous richness of the deposits, the Canadian authorities proposed a royalty of 20 per cent on the gold output, but there arose such a howl from the diggers that the impost was placed at 10 per cent on the gross value up to 1901, with an exemption of the first $5000. Later the exaction was reduced to an export tax of $2\frac{1}{2}$ per cent, which is now deducted by the banks when they purchase the gold.

The Government is generous, and wisely so. Foreigners have the same privileges as Canadians; no license to mine is necessary, every one having the rights of a 'free miner.' Not even during the corrupt period was there any discrimination against Americans, much as they feared it. On the whole, the administration of law and order is greatly to the credit of Ogilvie, of the other Commissioners who succeeded him, and of the Mounted Police, for all of them underwent the severest trial of all, namely, the vision of much gold and the chance to get rich quickly. We shall see that at Nome the temptation proved too much and that anarchy ensued.

The administration of the Yukon Territory is vested in a Council, headed by a Commissioner, who is the representative of the Dominion government. Six commissioners have held office since 1898, the first being Ogilvie. There is no specified term of office, the appointment coming at the hands of the Premier of Canada. The salary is $12,000, and extras. The Council is half elective and half appointive, the latter being usually local officials. As the Council can act only on the initiative of the Commissioner, and no money can be voted except at his suggestion, the Government at Ottawa is in practical control. It is not representative government, but it is more nearly that than the spoils system by which Alaska gets judges nominated by political bosses in Montana and North Dakota. The Commissioners have been men of character, even though

ON THE VALDEZ TRAIL IN WINTER.

some of them were occasionally bespattered with political mud. Crude 'graft' has been absent, but during election time the Commissioner has been known to increase the force on road-repairs in order to get votes. The Gold Commissioner—and, by the way, the title of Commissioner is worked to death—is the most important of the subordinate officers, for it is his duty to interpret the mining regulations, except in so far as appeal may be made to the courts. At one time this functionary held a court for petty offences connected with mining, the cases being decided forthwith on their merits without excessive technicalities. Now all such matters go to the Territorial Court, which becomes a Supreme Court when the three judges sit in full bench. The Gold Commissioner also is appointed at

Ottawa. Three men have held this office successively. In the first years, 1897 and 1898, there was crookedness among the subordinate officials in the Gold Commissioner's department, notably in the Recorder's office. Men having claims to record were refused grants because the ground was stated to be not open for location, although it actually was vacant, but it was desired to give the friends or accomplices of the Recorder a chance to locate first. It was the custom also to give women the right of way at the Recorder's wicket, and this led to collusion, particularly through the agency of the loose women who then infested the town. Mass meetings were held to voice the public indignation, and as soon as communication was established with Ottawa these wrongs were righted, the Gold Commissioner then in power was removed, and such performances have not been repeated. At no other time was there a real break-down in the system of local government.

The judges are appointed for life and receive good salaries, from $10,000 to $12,000 a year, besides perquisites. Men of high character are selected and hold office during good behavior. The prosecuting attorney or Crown Prosecutor is appointed by the Dominion Government, for life and good behavior. He is apt to be one of the leading legal practitioners in the locality. He is paid by fees, not by salary, rendering his bill to the Department of Justice at Ottawa. In an American mining district the District Attorney is usually a small politician, and the people are likely to be represented by a second-rate lawyer, while the criminal engages the cleverest member of the profession. The Territorial Judges are paid by the Department of Justice at Ottawa; thus they are independent of local sentiment.

CHAPTER XXII.

THE YUKON DITCH.

The most important work of engineering connected with the exploitation of the golden gravels of the Klondike and its tributary creeks is the construction of the Yukon ditch. This system of pipe, ditch, and flume has a total length of 70 miles. It was my good fortune to observe the building of this conduit, which brings the turbulent waters of the Tombstone river over hill and vale to the diggings near Dawson. In company with Messrs. O. B. Perry, C. A. Thomas, Scott Turner, and W. F. Copeland, I rode to a number of points, affording examples of types of construction, difficulties of the work, and the character of the life in camp.

Several schemes for bringing water under pressure to the placer mines on Bonanza and Hunker creeks have been considered during the last five years. One of these involved the use of the water flowing in the Klondike river, but it was ascertained by survey that the low gradient of that stream would necessitate a ditch fully 85 miles long and an expenditure of about $7,000,000. A. N. C. Treadgold, the promoter of the enterprise now known as the Yukon Gold Co., made surveys along the tributary streams flowing into the Klondike and the Yukon from the north. Finally, he applied for a right of way for a ditch to tap the head of the Twelve-Mile river. This enters the Yukon 18 miles below Dawson, and has its source in the Tombstone range, a part of the Ogilvie mountains, which rise to an altitude of 7000 feet, and gather sufficient snow to furnish a constant supply of water. It was estimated that a ditch and pipe-line to the mines near Dawson, with a capacity of 125 cubic feet per second, delivering water under a head of

350 feet, would be 70 miles long and would cost $3,000,000. It has cost over this amount up to the present time, and will require a further expenditure of $500,000. The total distance between the head of the ditch and Gold hill, the point of distribution, is 70.2 miles, the difference in elevation between these points being 1112.8 feet. The effective head along Bonanza creek, in the vicinity of Gold hill, is 375 feet. The construction includes 19.6 miles of flume, 38 miles of ditch, and 12.6 miles of pipe. Owing to the nature of the ground traversed, it has been necessary to modify the size and gradient of the ditch according to local conditions, but the standard is a 9-foot bottom, with 3½-foot depth of water, and a gradient of 6 feet per mile, ranging from a minimum of 4 to a maximum of 7 feet per mile. In places the ditch is fully 20 feet wide. The standard flume is 6 feet wide and 4 feet deep, with a gradient of 0.2841% or 14 feet per mile. The pipe varies according to the engineering requirements and is variously built of steel and wooden staves, so as to have a diameter ranging from 42 to 54 inches.

Wherever practicable the water is conducted by ditch, for that is the cheapest and most durable conduit. A ditch is necessarily dependent upon the contour of the surface; where depressions exist, a long detour is saved by building either a flume or pipe. If the depresison is a deep ravine or a broad valley, it becomes impossible to construct a flume, and recourse is had to a pipe in U-form (forming a so-called inverted siphon), the loss of effective head being measured by the friction between the water and the sides of the pipe.

The country traversed by this ditch is a rolling woodland indented by the alluvial flats of the Klondike, the Twelve-Mile, and other streams flowing into the Yukon river. As seen from a height the wilderness stretches unbroken from the meandering shimmer of the Klondike, enclosed within high banks on which white scars mark bench-diggings, to the Ogilvie range, where, far to the north, the snow still lingers in token of the gift of water that shall enable man to win the gold from the deposits of gravel strewing the tortuous valleys. The engineer who first planned the line of flume, ditch, and pipe had that kind of constructive imagination which is the creative force behind all engineering work. He imagined the deed done, and then he calmly began to calculate how to accomplish it.

THE TOMBSTONE RIVER AT THE INTAKE.

As viewed from afar the panorama of wooded valleys and the distant ranges serving as a water-shed, afford no suggestion of the natural obstacles to be overcome, but a closer acquaintance soon demonstrates that the forest is but a scant growth of small trees, just fit for telephone poles, not big enough to yield lumber, struggling to assert a stunted life amid the vast morass covering the face of the land. A wet blanket of moss mantles the ground, which is held in the grasp of a perpetual frost. Under the moss is ice; the moss forms an insulating blanket so that even the short warm summer does not thaw the frozen ground lying beneath this dark green coverlet. In places the ice melts slightly and pools of water form. Everywhere the surface is wet and sloppy. Our horses splashed through it. We stumbled over the spongy mass. It is a dismal swamp, which becomes almost impassable when torn by traffic. Wherever a trail was worn by use, it became a quagmire, and it was best to turn our horses to the untrodden moss alongside; in this their feet would sink only to a depth of 6 or 8 inches, for below that was the frozen ground; but where, in places, the moss was cut and worn away, the thaw had reached deep enough to make progress impossible. And these conditions obtained not only on the flats, but on the slopes. The water is held by the moss, so that even over an undulating topography there are no running streams.

In returning, Scott Turner and I took a short cut, leaving the main road and following a bridle path that ran across country by the side of the 'pole-line', that is, the wire of electric transmission. The trail ascended a hill that would be deemed steep under any conditions, but being covered with wet moss it seemed like the side of a house. The tundra had been trodden by the hoofs of horses and was soggy, pools of water enclosed by moss hung to the slanting surface, and when half-way it seemed as foolish to advance as to retreat. Fortunately neither the horse nor the rider laboriously leading him sank beyond a limited depth, which was fixed by the underlying frozen ground, so that a footing was obtained even in the worst spots. The incident is related in order to emphasize the fact that the steepness of hillsides does not suffice to drain the ground, the water from the thawed surface being held as by a sponge within the covering of moss.

In preparing to build the ditch, the first step was to place a saw-mill on the Twelve-Mile river, and thus to obtain the lumber for construction. Then an electric generating plant was erected, and the wires were strung on poles for 36 miles, transmitting power from the Little Twelve-Mile river to Bonanza creek. While this was being done, surveys for the ditch were hastened. As soon as the surveys were completed, the right-of-way was cleared. The small growth of forest was removed, and the moss stripped from the frozen ground for a width of one chain (22 yards). Then steam-shovels were put to work, and while they were digging the ditch, the saw-mill on the Twelve-Mile yielded the lumber needed for the construction of the flume and for other purposes. Seven million feet (board

FINISHING THE DITCH.

measure) of lumber was cut: this depleted the small forest in the vicinity, but it proved sufficient.

Without the steam-shovel it would have been hardly possible to dig the ditch in an economical manner, for manual labor at $4 per day, plus board at $2, or a total of $6 per day, is a costly instrument of engineering. Six shovels were employed. These made the cut, which was then beveled by hand, to be followed by the laying of moss on the sloping sides, with a little fine dirt as a finishing touch.

Roads of the corduroy type have been constructed, moss being laid on the poles and dirt on the moss. The trails traverse the brush in straight lines. Horses and men, steam and muscle, have fought against the wilderness and subdued it. The big ditch looks like a Panama canal, and the steam-shovels groaning and digging in the deep cuts recall pictures of Culebra. Many of the laborers had worked on the Isthmian canal, and assuredly the young engineers were as proud of the work they were accomplishing as if it were a national or even an international enterprise. The wilderness that had lain in shivering silence for untold ages, responsive only to the footfall of the moose and the caribou, hearing only the voice of the stream and the crash of the tempest, has been invaded to the very threshold of the Arctic by insistent man, determined to use Nature to his purpose, to overcome her obstacles by turning her own energy and her own power to his good in the quest for gold.

While riding to various points of interest and noting the progress of the work, I had an excellent opportunity of viewing this Northern land as it looks before it is scarred and disfigured by the encroachments of civilization. The dominant feature is spaciousness; it is a land of magnificent distances. Standing on the top of any of the higher ridges the landscape is impressive rather than beautiful, and splendid because it is vast; the shadows of swiftly moving clouds sweep over the green slopes, the air is still, the undulating forest is unruffled, and a great repose broods over nature as at the dawn of creation. It is a primeval wilderness; man and his handiwork are lost in the immensity of the setting, the only suggestion of his invasion being the white dots of tents, the slight scar of the ditch belting a hill, and the undeviating line of poles carrying the wire wherewith the engineer harnesses the torrent to his bidding. There is no sound; every footfall is muffled by the moss; no birds sing, no insects are heard, even the predatory mosquito attacks in silence. That multiplicity of life, of insect, bird, and beast, which makes the tropics intensely alive, is absent in the region bordering the Arctic Circle. Only at the noon hour or in the evening we heard the whistle of a steam-shovel, its little call to rest and food being instantly swallowed in the vast stillness. There was no echo. Approaching Lepine creek where pipe-joints were being made, there came a cheerful

ASSEMBLING STAVE-PIPE ON THE LINE OF THE YUKON DITCH.

sound of crimping and rivetting, reminding the citizen of San Francisco of the re-building of his ruined city and of Charles Field's lines:

"From mighty roots of concrete deep
The giant flowers spring from sleep
Along the barren highways of the city of my love."

Our short sojourns at the camps were most pleasant; the food was excellent and the keen interest taken by the members of the engineer staff in anything relating to their work, or to mining generally, prompted many interesting conversations. The men in charge were young, the chiefs of divisions being about 30, even the general manager and consulting engineer had only seen 33 years, and anyone over 40 was rare. Most of these intelligent and energetic young fellows were graduates of universities and schools of mines; the consulting engineer came from Columbia, the manager from Stanford; among other universities Harvard, California, and Michigan were represented; if Stanford had the lion's share of appointments no one complained, for the manager was a famous football player from Palo Alto.

Among many pleasant evenings I remember best the one spent at the camp on Lepine ridge. The sun set in golden curtains, and day lingered so long that at 10 o'clock we were still playing quoits with horseshoes. When it was time to sleep it still seemed too early despite the exertions of the day. The camp commanded a fine view, looking toward the Twelve-Mile valley and the Ogilvie range. In the west the clouds are still edged with glowing light; against this background the mountains stand silhouetted in sculptured masses of deep purple; in the middle distance stretches a trackless expanse of rolling woodland from which rises a thin wreath of smoke—not the incense from an altar, but the exhaust of a steam-shovel, marking the farthest outpost of industrial conquest, and near it is seen the line of pipe and ditch, becoming a mere thread in the gathering twilight. As the brief night approaches even these traces of man's doings are lost in the boundless wild, and the scene loses a jarring note. A pearly light suffuses the

WOODEN STAVE PIPE-LINE.

PIPE CROSSING THE KLONDIKE.

wilderness, the band of clear amber on the horizon fades into the pale flawless blue of the arching sky, in which even now no stars are visible. In the farthest distance the serrated mountains are shrouded in a night mist, and from the valley comes the faint voice of the river. Peaceful and vast is the picture. Thus it must have been at the beginning when the Creator brooded over His handiwork, before the footfall of man's oncoming had been heard, before bird sang or flower grew, while as yet the earth was fresh from the making, and strife had not begun.

CHAPTER XXIII.

FROM DAWSON TO FAIRBANKS.

On July 28 at 11 p.m. we left Dawson on the *Lavelle Young*, and in a few minutes we were gliding northward on the full flood of the Yukon, whose silvery waters caught the glow of a twilight that joined the beauty of the sunset with the promise of the dawn. The next morning, just beyond Forty-Mile, we crossed into Alaska, the boundary being indicated by a clearing 20 feet wide cut straight through the bush with undeviating line. We were once more in American territory. The scenery continued monotonous: on one side a headland; on the other, flat high bank and low shore alternated as the river followed its sinuous channel to the sea. Eagle, one of the pioneer settlements, is interesting on account of its association with Jack McQuesten, a hero of the days when starvation threatened the widely scattered band of adventurers in the northern wild. The old store, marked McQuesten & Co., belongs to the Northern Commercial Co., and over it looms tall and spare the tower of the wireless telegraph station, indicating that this, one of the oldest mining camps in Alaska, is at last in touch with the outside world. Here an Army post has been established. The next port of call was Circle City, another old settlement, deriving its name from the supposition that it stood on the Arctic Circle, which is still several miles north. Just below Circle the Yukon widens into the Flats, a dreary stretch of sand-bars and swampy islands.

Early next morning we crossed and re-crossed the Arctic Circle, that imaginary line which runs around the earth at the latitude of 66° 33′ north. At 8 a.m. on July 30 we called at Fort Yukon, which is 8 miles north of the Arctic Circle, and I wrote a few letters to friends to tell them that "we change our skies but not our hearts." The settlement now called Fort Yukon is a few miles below the old Hudson's Bay post, founded in 1847; it is a river station for those going to the Koyukuk mining district, the centre of which is Bettles, 150 miles by trail northward. Below this "farthest north" of the Yukon, the river swings southwestward and opens into a continuation of the Flats—a monotonous scenery. The smoke from forest fires smothered the landscape, and we were not surprised to pass a new settlement labeled Purgatory.

The voyage is punctuated by frequent halts for 'wooding-up,' the term given to the process whereby fuel is transferred from shore to boiler-room. Thus two hours are consumed each day, for the *Lavelle Young* burns 26 cords of wood in 24 hours. She is 140 feet long, 32 feet in the beam, and draws 3 feet of water. The current of the Yukon between Dawson and the Tanana averages 4½ miles per hour and by the aid of it the steamer covers 15 to 16 miles per hour. From Dawson to Forty-Mile the current attains a velocity of 6 miles per hour. Here are the distances and the time made by the *Lavelle Young*; someday they will be interesting, although today they are not:

	Miles.	Hours.
Dawson to Forty-Mile	54	3½
Forty-Mile to Eagle	54	3½
Eagle to Circle	165	10½
Circle to Fort Yukon	80	6
Fort Yukon to Rampart	225	13½
Rampart to Tanana	75	5¼

From the captain I learned that skilful navigation on the Yukon is unlike the kingdom of Heaven, for it "cometh by observation," and not by charts. The channel is changing continually, owing to the undermining of the banks by the ice and the shifting of the current. How rapidly the banks are undercut is indicated by the fact that wood-piles placed 50 to 60 feet from the edge are now, only two months later, close to the water, while on the opposite shore the river is depositing silt with compensating rapidity. At one place on the Yukon Flats the river had encroached 125 feet up to the end of July; at another place on this river about 50 miles below Tanana, columns of frozen earth fully 100 feet high are constantly being loosened from the bank; these tough frozen masses gradually incline forward as they become undermined by the current, until finally they fall into the water with a splash that would endanger even the largest boat on the river. On the Koyukuk these masses detached from the bank contain enough ice to float on the stream and enough soil to bear standing trees; thus they

ARRIVING AT EAGLE.

constitute veritable floating islands. Capt. Gray tells me that he has seen them as long as 150 feet, drawing 9 feet of water. Hence the particular difficulty of piloting vessels of even shallow draft in this part of the Yukon; hence also a special pilot comes aboard at Circle City, taking the boat in hand as far as Fort Yukon. To overcome the dangers due to shoals the steamers are built in the form of mere skimming dishes; for example, the *Koyukuk*, which runs 520 miles up the river of that name, is 24 feet wide and 120 feet long, she is made of inch boards on the sides and 1½-inch stuff on the bottom. When empty she draws 6 inches forward and 10 inches aft, when her boiler is full of water she draws 10 inches forward, and when loaded with 60 tons of freight her draft is 21 inches only. The *Lavelle Young*, on which we traveled, had a history: she was originally designed as a scow for a dredge which some Columbia river pilots purposed placing on the Koyukuk. Under her own steam she was taken from Portland to St. Michael, where a fancy price was offered for her in 1898, the dredging scheme being abandoned in favor of the assured gains of river traffic; thus the *Lavelle Young* was chartered for passenger and freight business on the rivers of Alaska. She has three main rudders each 12 feet long and two monkey-rudders each 8 feet long. No survey has been made of the river, so that all the pilot has to aid him is a track chart prepared by himself or the captain, recording the course with regard to prominent objects. The points on the river are named after the wood-yards or the camps of Indian traders. A slough will be christened with the name of a steamboat that made a mistake and roosted there for a month or so; sand-bars are named after boats that got into trouble upon them; thus a haphazard nomenclature is evolved.

At 1 p.m. of July 31 we reached the confluence of the Tanana and the Yukon, the meeting of river-trails being marked by a settlement having three names—Tanana, Fort Gibbon, and Weare. An Army post is here established, two companies of the Twenty-Second regiment being in quarters. The village stretches along the right bank of the Yukon and consists of many saloons, several large mercantile stores, and the barracks; at the northern end of the settlement a dozen tents along the river bank are occupied by prospectors, about to proceed up the Innoko, Koyukuk, and other streams offering a golden lure. It is a pity that the coming and going of these nomads is not recorded, together with information concerning their finds and the conditions observed on the various creeks; they seem to follow each other in a shiftless manner, one party going where only the season before another party had tested the ground without success. They suggest a pitiful waste of energy and time, as well as knowledge.

A dull day ended in a wonderful night, ushered by a resplendent sunset. The wretched little steamer moving to the irregular rhythm of the paddle-wheel at the stern was without a nautical curve in her design or a glimmer of poetry in her appearance, but when I stood in the pilot-house beside Capt. C. A. Boerner while he steered his vessel into the gateway of the sun, we entered a veritable realm of enchantment. The vessel is gliding forward on the swift broad current, no sound

CAMP ON THE INNOKO.

is heard save the ripple of the parted wave and the deep breathing of the engine; we are hastening into the sunset mirrored in the shining water, but we come no nearer to the splendor, which beckons ahead. A headland throws a pyramidal reflection across our path; we cross it with silver rippling; the low dim land on either side fringes the river with mystery, while far and faint a single mountain peak in the heart of the wild stands sentinel. An island is passed, and yet another, before the boat enters a broad sheet of radiant water. The sunset, still ahead, glorifies the way. Only a purple headland separates the splendor sailing in the sky from the beauty swimming in the wave. A wind, balmy as the breath of spring, is wafted from the south; the frozen land has surrendered to the stainless glory of the moment, and the ineffable beauty that looks down upon the sleeping wilderness.

They say that those who have lived more than a year within the interior of Alaska suffer a nostalgia that knows no cure save the return to the land they love; they feel the pull of a magnet that draws them from the busy haunts of men back to the lonesome outposts of the North. It is the love of the sailor for the sea, of the mountaineer for the peaks, of the prospector for the desert. If the briefest of sojourns in Alaska may permit of a true impression and a touch of sympathy with the homesickness of the Alaskan, I would impute the haunting charm of the country to the spaciousness, the untrammeled life, the freedom from restraint, the old ineradicable Adam that loves to be his own master as in the days before the serpent complicated life by suggestions of disquiet. Prospectors, miners, woodsmen, and explorers who come to the big cities soon find that they are nobody; the things they can do well avail them nothing; the multiplicity of things they cannot do shames them; from being monarchs of all they survey they are subjects of all whom they see. No wonder they yearn for the boundless horizon, the untainted oxygen, and the perfect emancipation of life on the far outposts of empire. To come to the relaxing atmosphere of civilization is depressing to one of these, for they have breathed an air as pure as inter-planetary space, they have felt the physical stimulation of the dawn "when the mountains were flushed as with wine, in the morning of Time."

At Fort Gibbon passengers are usually transferred to a smaller boat, which takes them up the Tanana river to Fairbanks. Navigation becomes more difficult, for this tributary of the Yukon is swift and uncertain; a shoal may exist today where adequate flotation was obtainable a week ago; therefore, we did not protest when the steamer ran aground, merely congratulating ourselves when we got off. At 7 p.m. we passed Nenana, a pretty name associated in my memory not with a lithe Indian maiden or a graceful canoe, but a curious fact in physiography. By ascending the Nenana river the traveler can reach a group of lakes on Broad pass, the divide separating the

STEAMER 'WHITE HORSE' ON THE YUKON.

watershed of the Tanana from that of the Susitna, whose waters flow into Cook's Inlet. Similarly, by following the Delta, another tributary of the Tanana, to its source, one can reach Gulkana lake and proceed down the Gulkana river, a tributary of the Copper river, into Prince William sound. Finally, by proceeding to the headwaters of the Tokio, a third tributary, the Mentasta lakes are reached, and from them it is possible to descend by Mentasta creek into the Salana, thence to the Copper river, and onward to the sea. These three instances show the lowness of the divide between the watersheds, for in each case a waterway is available by means of a lake linking a river on the east with another on the west. Other cases might be mentioned in the same region, where a short portage will transfer a canoe from one watershed to another. This is an unusual type of topography. The same condition is indicated by the sinuous course of the rivers; from Tolovana to Nenana it is 65 miles by river and 28 miles across country. "Two bends and a look" is a local mode of describing the distance between points.

When the *Lavelle Young* met the *Tanana*, the passengers for Fairbanks were transferred to the latter, much to their satisfaction; at the same time a number of mail-sacks containing gold were shifted, each sack holding 22 bars weighing 4 pounds apiece. They were treated with scant ceremony. The *Tanana* is an excellent little boat, ably commanded by Capt. James T. Gray; she is 150 feet long and 30 feet in the beam; when launched her draft was 14 inches, with steam up, but without equipment; now she draws 18 to 22 inches of water; when loaded with 150 tons her draft is 32 inches aft and 38 inches forward, each additional 10 tons lowering her an inch into the water. Her engines are horizontal, simple, non-condensing, with 13-inch cylinders and a stroke of 6 feet, the valve-motion being controlled by a link actuated by two eccentrics from the paddle-wheel shaft. The hull is coated with carbolineum. The barge she pushed was 100 feet long, had a draft of 39 inches, and carried 140 tons. In the management of this barge, Capt. Gray showed a skill which it was pleasant to watch. While the boats on the upper Yukon have a square nose or bow, the

STEAMER RECEIVING WOOD ON THE YUKON.

Tanana has the usual pointed bow, which impinges on the stern of the barge. By manipulating the ropes attaching the two vessels to each other, the bow of the steamer acts as a fulcrum when turning the barge in a sharp bend of the channel. This method is called 'jack-knifing.' The deflection of the barge is managed by means of rope and tackle, operated from a double-headed winch; the ropes being pulled in or let out on alternate sides so as to point the barge this way or that according to the bell signals of the pilot. While this is being done, two men (one on each side of the forward end of the barge) are sounding, and as they cannot be seen or heard by the pilot owing to the housing and distance, a third man stands on top and repeats the soundings to the pilot. By placing the barge ahead of the steamer, navigation in a tortuous channel is aided, for the barge acts as a large rudder, preventing a steamer of shallow draft from being swept around by the current. Note may be made of the fact that during summer in the upper reaches of the Tanana, there is sometimes a daily rise of as much as two feet, owing to the melting of glacier ice during the warm hours of the day, causing a river tide.

The many tributaries of the Yukon are navigable to their headwaters by steamers of not over two feet draft, the gradient of the streams averaging about one foot per mile despite occasional rapids. Thus small steamboats have ascended or descended the Teslin from the head of the lake, the Lewes from Lake Bennett, both the Porcupine and the Chandlar for 150 miles, the Tanana, for 700, the Koyukuk and the Innoko, for over 500 miles. All of which testifies to the low topography of the interior of Alaska and the Yukon.

Prospectors ascend the smaller streams by means of 'poling,' one man standing at the bow and another at the stern. The poles are 12 to 14 feet long, pointed with iron, and the boats are flat-bottomed for at least two-thirds of their total length. They draw from 7 to 8 inches and carry 1200 to 1500 pounds. By this method from 10 to 15 miles per day is traveled.

Between Fort Gibbon and Fairbanks, the current averages 4 miles per hour, with a maximum of 7. When pushing a barge, the *Tanana* made 6 to 6½ miles per hour against the stream and a speed of 20 miles coming down-stream, when she "ran like a scared dog." The fuel is wood, which is consumed at the rate of a cord per hour and $7 per cord. At Dawson, owing to the distance from uncut forests, the price of wood is $14; down the Yukon it is $7 to $8 per cord.

The little flat-bottomed steamers are well adapted to the shallow rivers of Alaska. One of the most useful was the first *Koyukuk*, which, with steam up, drew only 8 inches forward

POLING ON THE INNOKO.

and 10½ inches aft, when loaded with fuel for three hours, besides all her equipment, and stores. Her two engines were 9 by 48 inches. This friend of the prospector was wrecked on the upper reaches of the Tanana during a severe windstorm, such as she was ill fitted to meet on account of her lightness above water. Another boat of the same name has been built, and to her reference was made earlier in this chapter.

Meanwhile, we were progressing toward Fairbanks. At Sixteen-Mile the *Tanana* went over a place where only three years ago there stood a road-house—a reminder of the vagaries of stream-erosion. Each break-up of the ice in the spring starts the cutting of a new channel, according as the ice restrains or releases the accumulated water. Approaching Chena, we passed a refrigerator barge, suggesting the manner in which perishable foodstuffs are transported to the interior of the country. This barge was towed from Seattle to St. Michael, where she received her cargo, 300 tons of meat and poultry, worth $200,000. At Fairbanks the retail price is 40 to 50 cents per pound. As we passed, the ill regulated explosions of a gasoline engine indicated the power used for refrigeration. Soon after, we met a man in a punt, using a single pole. When going up-stream, keeping to the shallows, he would use two poles, one in each hand. From his boat we saw smoke arising as if from a miniature boiler; it was a 'smudge' or dank fire to ward off mosquitoes.

At noon on August 2 we 'hove to' at Chena, where the boats leave the main stream and ascend a slough ten miles to Fairbanks. Approaching the town, log-cabins come into view dominated by the tower of a large school-house, proving that a thirst for knowledge is encouraged in the very heart of the Alaskan moorland. Another large wooden structure proves to be a brewery, the evidence of thirst of another kind. On the outskirts of the settlement, the gardens of potatoes and cabbage indicate the beginnings of agriculture and a protest against canned vegetables, the use of which is suggested by large heaps of discarded 'tins.' Who said there was no tin in Alaska? More tin has been imported thither than has been produced in the mines. Capt. Gray told me that the *St. Paul* had taken 10 tons of tin ore in one shipment from Nome, and, said he: "Ten tons goes a long way." Truly; but not so far as the scrapped tin-plate has already come. The empty tin cans used to be thrown into the river, where they became filled with sand and impeded navigation. Now there is a United States law against the throwing of tin cans into the rivers; in time there will be a tin mine in Alaska.

At 3:30 p.m. we stepped ashore at Fairbanks and within an hour we were careering up the valley in a motor-car running on the rails of the Tanana Valley Railroad. Our host was Falcon Joslin, the president of the railroad, a gentleman pos-

LEAVING CAMP.

sessed of both humor and knowledge. We ran up to Gilmore (21 miles) and back, the return being made at a speed trying to the nerves of a 'cheechako.' Thus first impressions were gained and an appetite acquired. As to the latter, it was not of hopeless yearning, for did we not read in a store window: "Cracked eggs, 75 cents per dozen." You cannot make an omelette without breaking eggs; here then there was promise of an omelette. But we fared even better, dining extravagantly on perfect grayling and a blueberry pie made with the wild fruit of the Tanana moors. After 'supper' in the clear light of the Alaskan summer evening, we strolled through the town, noting the banners of rival candidates and signs indicating the headquarters of political parties in an election for delegate to Congress. Pretty log-cabins, with gardens full of flowers, lined the river banks and suggested the presence of the plucky women who accompany their mates in the search for gold.

CHAPTER XXIV.

FAIRBANKS.

The story of the founding of Fairbanks is one of the romances of the North; and like such romances it has its sordid page. In 1897 E. T. Barnette brought a boat up the Yukon, from St. Michael to Circle City, and thence to Dawson. In the spring of '98 he went 'outside,' returning to Seattle for awhile. The two following years were spent at Dawson. In the spring of 1901 Capt. Barnette, as he was now called by reason of having commanded a boat, bought a stock of goods, weighing 100 tons, at San Francisco; these he shipped to St. Michael, where he chartered the *Lavelle Young*, intended originally for a dredge-scow, and with her he steamed up the Yukon and the Tanana. His intention was to start a store where the Valdez and Eagle trail crosses the Tanana, at a point about 300 miles above the present site of Fairbanks. It was generally known that a belt of copper veins existed at the headwaters of the Tanana and White rivers. With this in view, Barnette went up the Tanana, on the *Lavelle Young*, which he had placed in charge of a man who proved unaccustomed to the navigation of swift and shallow rivers. Discovering that the boat had not sufficient power to go farther without the aid of lines for towing, he went 14 miles above the junction of the Chena to the foot of the Bates rapids. Unable to proceed farther, Barnette and his expedition returned to the mouth of the Chena and worked their way up the slough to the place where Fairbanks now stands. The site was chosen by reason of the high bank, promising escape from floods, and an unusual growth of forest, including many spruce 24 to 30 inches in diameter. Here they landed on August 24, 1901.

Another version of this part of the story is that Barnette did not try to go above Chena on the main stream of the Tanana, as it was then supposed that the Bates rapids began just at Chena. He went up the slough a few miles above the present site of Fairbanks, and finding further progress impracticable, he dropped back, in perplexity as to what to do next. He tied up for the night to a convenient shore and the next morning unloaded his goods. The master of the boat, Thomas Bruce, claimed that he had fulfilled his contract, which was to take Barnette to Mentasta crossing or "as far up as practicable." Thus the position of Fairbanks was fixed.

The steamer returned to St. Michael, leaving five men, besides Capt. Barnette and his courageous wife. Four of the party went out prospecting, and, of these, Dan McCarty located the claims known later as No. 1 Above on Fairbanks creek, No. 2 Below on Cleary, and No. 5 on Gold Stream.

Even before these locations were made, in the fall of 1901, a party of five men had come across country from Circle City, which is 175 miles northeast. Felix Pedro, Tom Jurack, Bert Johnson, and two others were prospecting for a creek that Pedro had discovered three years earlier. He had wandered into this region from the Forty-Mile, and while alone he had run out of supplies, lost his way, and finally reached Circle. In the course of his wanderings he found a creek containing gold; that creek has never been re-discovered and remains one of the 'lost mines' of the prospector's mythology. Knowing that Barnette was coming up the Tanana with supplies, the five men led by Pedro took the risk of exploring ahead of him. From a ridge east of the river, from a height now called the Pedro Dome, they saw the *Lavelle Young* coming up the Tanana. On the first night following the landing of Barnette, these men came into his camp. They were out of supplies and had been living on berries and game. Soon after it became generally known at Circle City that a trading post had been established on the Tanana and 30 men came across country in March 1902. On July 27, of the same year, Pedro made the first discovery of gold in the district at the head of Gold Stream where it narrows into Pedro creek. He picked up some gold in the bed of

THE FIRST CAMP AT FAIRBANKS; IN 1903.

the living stream and sank a hole into the frozen ground to one side. The discovery was kept secret among the dozen men then in the vicinity.

In the meantime Barnette was absent. He went 'out' in the spring of 1902, going by the trail to Valdez, thence to Seattle, where he built the *Isabelle.* This vessel was 'knocked down,' or disjointed, and shipped to St. Michael, where she was put together. Barnette arrived at his trading post on September 5, 1902, intending to proceed up the river, as originally planned. He now had a boat of light draft, suitable for his purpose. But, on arrival, he heard of the discovery of gold, and that, of course, changed everything. He decided to stay where he was and himself located several claims, others having been previously staked for him by that legal fiction known as 'power of attorney'—one of the curses of Alaska.

Among those who came with Barnette on the *Isabelle* was Wadda, a Japanese, a redoubtable character, who was then serving as cook, but intended to trade on his own account with the Indians. In January 1903, he went with a dog team to Circle and then to Dawson, where he told the news of the discovery made by Pedro and McCarty on Gold Stream. Both of these finds looked promising, for McCarty was getting as much as 20 cents per pan. At Dawson, Wadda met Capt. H. H. Norwood, a whaler with whom he had served in the Arctic; he urged Norwood to put up the money required to buy claims in the new district. Norwood proved unwilling, but in the meanwhile Wadda's talk had leaked out. A stampede ensued at once. In February, March, and April 800 men left Dawson for Gold Stream. Most of these were not real miners, but the floating population of Dawson; such men are called 'stampeders.' Arriving, they were soon disgusted by the absence of gambling and other gaieties; they were disappointed in seeing so little gold, forgetting in their hurry for fortune that it took time and work to produce an output. Thereupon they asserted that Wadda had deceived them, and wanted to lynch him. They also threatened an attack on Barnette's store; but a few well armed men thwarted their purpose. Subsequently, several hundred of these men left the district for the mouth of the

Yukon, departing in small boats and even on rafts. These fellows condemned the region, so that the supplies that had been intended for traders at Fairbanks were either diverted to other points or countermanded altogether. In consequence, a severe shortage of food was experienced during the ensuing winter, and many would have died from starvation, if, fortunately, great herds of caribou had not crossed the district on their annual migration southward, and if the supply of ptarmigan and rabbits had not been unusually plentiful. Soon after, Cleary creek was discovered and there was rich ground for everyone.

With the first snow that fell in the fall of 1904—in October—a boiler of 40 horse-power was hauled to Cleary and set up on No. 1 Below. This marked the beginning of real mining. Previous to that event the work had consisted mainly of "sniping around with little porcupine boilers," that is, desultory digging by means of thawing effected with crude appliances. The men first on the ground had ascertained that the 'pay' was deep and being themselves too timid to get the requisite machinery "they sat around and whittled" until the experienced miners from Dawson arrived and bought the claims.

A settlement sprang up around the camp built by Capt. Barnette and to the town was given the name of Fairbanks, in honor of the Senator who became Vice-President of the United States. In 1903 the population was about 800 and $35,000 in gold was produced; in 1904 the population increased to 3000 and the gold production to $350,000; in 1905, with a population of 6000, the yield of gold reached $3,750,000; and in 1906, with 8000, the output was $9,175,000. At the present time the town has 3500 people and the district about 15,000. In 1908 the gold output, despite labor troubles, was $9,250,000.

Fairbanks is an attractive settlement and an important distributing point. Someday it will be on a transcontinental railroad. In regard to this, I was enlightened by Mr. Joslin, an optimist restrained by good judgment. Alaska needs a trunk railway system and eventually such a system will become part of a line of communication between New York and Paris. It is suggested that a railroad should be built from Haines to Fairbanks, 650 miles, and thence to Nome, another 600 miles. Haines is at the head of Lynn Canal, just south of Skagway; it is a military post and was long known as Haines Mission. From Pyramid Harbor, near Haines, many prospectors in the early years started for the interior and bands of cattle were driven through the Chilkat pass along the Dalton trail to Fort Selkirk and Dawson. Over the neighboring passes the Indians of the interior used to bring the peltry sold to the agents of the Hudson's Bay Company in the days before Alaska belonged to the United States. The Chilkat pass is 2800 feet above tide-water and 50 miles from shore, so that at no place would the railroad grade be more than 2 per cent. North of the coast range the line would enter the Alsek valley, which is part of a broad ancient valley (the Shakwak) partly occupied now by the Kluane and other lakes. This district is suitable for agriculture, and, though but little explored, it is known to contain deposits of gold and copper ore, as well as coal. Then the railroad would cross the head of the White river, where are large copper deposits; thence over a grassy plain to the head of the Tanana and down that river to Fairbanks. No elevation has to be surmounted except in crossing the coast range at the Chilkat pass. Near Haines are copper, gold, and iron deposits. Twelve miles from this deposit, and 32 miles from Haines, are the Porcupine placers, in the Yukon, and beyond them in Canadian territory are the Alsek diggings. Coal seams are known to exist on this route at several places. One objection to such a railroad is the necessity for traversing Canadian territory for a short distance, but this is not a serious obstacle. As a steamer service all the year round is maintained even now with Haines, and as vessels can keep within the quiet waters of the 'inland passage' all the way from Seattle, this would furnish an excellent line of communication. Fairbanks is now 21 days from the 'States'; by building a road through the centre of the interior country Alaska would practically cease to be a "non-contiguous possession," becoming an integral portion of the United States. Two thousand miles of railroad, at $30,000 per mile, would accomplish this purpose.

Meanwhile we had the advantage of traveling on a railroad

that marked the beginning of improved communication, and as the train of the Tanana Valley railway system glided from the siding that constituted the Fairbanks terminus, we could without cost imagine ourselves on the way from New York to Paris, including a tunnel transit under Bering strait.

The Tanana Valley railway is ballasted with three feet of moss, which does not prevent the road-bed from being rough, in places. The railway gives convenient access to all the centres of mining activity. On the afternoon of August 3 we went to Chatanika, proceeding up the valley of Gold Stream, where men have dug through thirty or forty feet of frozen 'muck' and 'wash' to the bedrock on which the gold lies. The course of the former river-bed is marked by heaps of gravel, shaft-houses, and flumes. From the distribution of activity it can be seen how the gold-bearing channel traverses the present valley but follows a line independent of the topography of today. The train passes close to some of the diggings; men can be seen hoisting buckets of 'dirt' with a windlass and discharging them into sluice-boxes, where a flow of water separates the gold from the gravel. The valley of Gold Stream is wide and shallow, a depression in an undulating country, covered with moss. Several lakes diversify the morass, through which flows a meandering stream, with no more dignity than an abandoned and overgrown irrigation ditch. The scenery is not impressive; it is oppressive; yet as the train laboriously puffs its way up a ridge giving a view of Vault creek, the outlook becomes more cheerful. The white tents of the miners, and the big brown dumps adjacent, look like the anthills of a larger growth, and on the hillside beyond the lines of birchwood cut for fuel simulate the furrows of the farmer, suggesting cultivation rather than devastation. The contours are soft, the hills are velvety, the surface is dark and sodden, the sky is gray, and man and his doings seem but an insignificant irruption in this vast wilderness.

As it was only 6 p.m. when we left the train at its terminus in the village of Chatanika, we decided to walk to Cleary, a distance of three miles. After 'supper' a traveling show offered amusement. It was a biograph entertainment, given by a man, his wife, and young daughter. The scenes depicted on the screen were described by the showman, arousing keen interest among the miners and their friends, for the glimpses of a seaside resort, sail-boats, hotels, and children at play on the beach offered pictures of a life wholly different to the business of the 'creeks.' At the door of the hall, instead of carriages awaiting the play-goers, were the 'huskies' or native dogs, crouching in wait for their masters.

Next day, having visited three different mines on Cleary creek, we returned down the valley in the sunny afternoon, to Chatanika. The evening smoke is enjoyed while sitting on a bench in front of the hotel, watching the manifestations of local activity. Men come to the blacksmith opposite with 'points' that need repair and miners gather to discuss politics. Five candidates have been nominated for delegate to Congress and five newspapers are disseminating misinformation throughout the Tanana region. These papers sell for 25 cents apiece and they represent the minimum return for the expenditure. It is not surprising that Judge Wickersham, the man who received the least support from the local press, was elected. But the pitiful politics of the day are easily forgotten while watching the long shadows creep over the hillsides and the opal tints of the long Alaskan twilight wrap the hideousness of a mining camp in a glamor of mystical beauty.

On our return by train to Fairbanks, the passengers included one or two men who came on board at intermediate stations, carrying heavily laden common sacks which they placed to one side casually. Inside these was a leather 'poke' or bag containing the gold 'dust' from the clean-up of the day before.

CHAPTER XXV.

CLEARY CREEK.

On Cleary creek the traveler may see examples of the type of mining most characteristic of Alaska. The method is called 'drifting', because it is based upon the driving of a 'drift' or underground gallery from the bottom of a shaft. Under the existing surface of the valley lies the former bed of the creek, and on the rock over which the water used to run now lies the gold; the precious metal has been concentrated by reason of its greater specific gravity, causing it to sink through the débris of the stream to the bottom. Later, this old channel, with its gold-bearing sediment, has been covered by gravel and by 'muck', both of which, being wet, have been frozen solid. This freezing took place during a period when the cold of winter exceeded the thawing in summer, and, by reason of the subsequent growth of an insulating covering of moss, even the warmer climate of the present geological epoch is unable to melt the mass of material that lies on top of the old stream-bed. The creek now flowing in the valley is a rivulet, barren of gold, and otherwise of no consequence, except in so far as its living water has thawed the ground over which it courses. Therefore the miner avoids sinking his shaft along the present stream; he goes to one side and penetrates through the frozen ground to the old creek-bottom.

It will be best to illustrate the method by describing a specific mine, for instance, the one known as No. 11 Below Discovery on Cleary. The shaft is 7 feet square, and is 70 feet deep; in order to get a well for collecting the drainage, the shaft is sunk 8 or 10 feet below the surface of the bedrock. Then a 'level' or 'drift', 6 feet wide and 6½ feet high, is run up-stream along the top of the bedrock for a distance of 200 feet to the boundary of the claim. This main 'drift' or gangway is timbered to protect the miners and to prevent falling dirt from blocking the passage. Having completed the 'drift', similar galleries are run at right angles on both sides to the limit of the gold-bearing sediment, in this case 240 to 300 feet wide. Now begins the excavation of the deposit. To thaw the frozen gravel, 'steam points' are used, as already described in connection with mining at Dawson. The 'points' are first fed with hot water while they are being driven into the ground, and when they have been forced to their full length, steam is turned on for 24 to 30 hours. This is the 'sweating' period, during which the ground is thawed around each point for 2 to $2\frac{1}{2}$ feet. When the gravel has been softened, the points are withdrawn and the miners use their picks to break the material so that it can be shoveled into wheelbarrows, which are then trundled to the shaft. There the 'dirt' is discharged into a bucket, which is hoisted by a steam-engine to the surface, for washing in sluice-boxes. About one foot deep of bedrock and 6 to 7 feet of overlying gravel are removed by these mining operations. The excavation of the top of the bedrock is made necessary by the fact that the gold has sunk into the cracks and crevices of the rock, which, however, is so decomposed usually as to be soft and easy to dig.

We went underground, standing erect on the edge of the bucket and holding the steel rope, while being quickly lowered to the bottom of the shaft. Lighting the candles offered by the manager, we walked along the boarded way over which the wheelbarrows pass. Being warned of their approach, we stepped to one side with our backs against the side of the level, while the procession of six men trundling wheelbarrows proceeded toward the shaft. Each barrow holds 15 cubic feet or 375 pounds of gravel, six of them being enough to fill the bucket. The men are paid $5 and their board, which is worth $3 more per day, so that each laborer costs the mine-owner $8 per day. Each man picks, shovels, and wheels his own share of the output of the mine at a pace regulated by the leader, who is chosen by the manager. Turning to one side we entered a cross-drift communicating with a low cavernous chamber made by the removal of the gravel in the course of mining. There we saw a group of 38 'points' silently at work, with nothing to indicate the process, for all leaks of steam are carefully prevented. Such leakage not only means waste of energy but leads to heating the air in the mine and the consequent thawing of the roof of the workings. Everything is in a frozen condition. The air has the feel of a cold-storage chamber. In walking through the workings one hears the dropping of gravel loosened over-

LOWER CLEARY CREEK IN 1907.

head by the slight warming of the air by the bodies of the miners and by the little heat given out by the steam-pipes. Occasionally the visitor will not only hear, but feel, the crumbling of the over-arching ground, for a chunk of 'dirt', obedient to the immutable law of gravity, will tap him on the shoulder. To avoid danger, it is best to keep close to the frozen sides of the excavation, avoiding a position under an overhanging stretch of gravel. Returning to the main gangway, we crossed to the other side of the mine, where the men were removing the gravel previously softened by a battery of steam-points such as we had seen at work. No time was being wasted. Nowhere have I seen men working more efficiently, and although their wages are high, it must be remembered that the cost of living is abnormal.

By examining the sediment on the bedrock it is possible to see the yellow-red particles of gold. In a country of cheap pauperized labor, such as Mexico, these nuggets would be purloined by the workers. A high price for labor and supplies tends to depreciate gold; at Fairbanks a nugget weighing half an ounce represents only one day's pay.

The mines are worked mainly by lessees; fully three quarters of the gold extracted in the Fairbanks district is taken out of the ground by 'lay-men', who pay the owners of the claims a royalty of 25 to 50 per cent on the gross output. For this rich tribute the claim-holder has usually done nothing beyond locating the ground or having had it located for him. This is a striking example of the unearned increment and of special privilege under a democracy. Of the hundreds of rich owners in the district, only a few made a discovery of gold themselves, and only a few ever did any real work on their claims. Many of the claims were located under 'power of attorney', and in some instances by abuse of this privilege a few men have been able to acquire large areas of profitable ground. A miner can go up a creek and stake for himself and his friends as many claims as he likes, provided he makes a discovery on each 20 acre claim; although, as a matter of fact, a true discovery is rarely made, for the gold is thirty or forty feet underground, and the shaft to bedrock is not sunk until many days after the legal routine of 'location' has been completed. Moreover, a man can stake an 'association' claim of 160 acres, using the names of eight men and making only one supposititious discovery. By virtue of 'power of attorney' he uses any names he pleases, the fiction involving no permission or legal authority from the owner of the name. In this way Alaska has been blanketed with claims now belonging to men who sit back, letting others do the work and incur the risks of mining while they gather a rich tribute.

In April 1907 there was a strike among the miners in the

CLEARY CREEK IN WINTER.

Fairbanks district, and in the spring of 1908 this strike was resumed. At the time of my visit it was nominally in force, but the back of it was broken. After visiting the mines, seeing the conditions under which labor is performed, noting the character of the work, and the system of ownership, operation, and exploitation, I found myself in sympathy with the laboring men. This does not mean any sort of sympathy with the ignorant politicians who led them, nor the reptile press that incited them to reprisals. If one judged the cause by its advocates one would condemn the labor movement in Alaska in short order, but in justice to the men the following facts must be stated: The cost of coming to the mines is high; thus, by winter trail to Fairbanks from Valdez, 376 miles, the fare is $150 and the road-house expenses average $6 to $7 per day for 10 days. It costs $250 to come from Seattle or San Francisco. While the mine-owners provide board and bunk, the season lasts only 4 to 5 months, for at the first approach of winter all surface work ceases and half the laboring population is forced into idleness. The work in the drift mines is exceedingly hard; a man will average from 80 to 100 wheelbarrows, equivalent to 6 or 7 cubic yards, per day, and it needs an engineer to appreciate what that means. Suffice it to say, that it represents the maximum of manual labor. Then be it remembered these men are working for owners and operators of their own kind and class; in many cases nothing separates the sturdy miner trundling a heavily loaded wheelbarrow from the idle man who owns the mine, except an accident. A few months hence the positions of the two men may be reversed; not education, not ability, not technical knowledge, distinguishes the owner of a bonanza claim from his employees—simply chance and a mining law of absurd laxity. Of course, $5 per day, with $3 worth of board besides, looks large wages, but it is only $125 to $150 per month for four months, and from it must be deducted $300 to $500 for coming and going to the States, plus incidental expenses while at the mines. Here is another trouble; the miner pays 25 cents for a glass of beer, 50 cents for a magazine, 25 cents for a wretched squib of a daily paper, and so forth. And he needs diversion; a man under laborious exertion for 10 hours each day is entitled to relax and to rest, both mentally and physically. The cure for the abnormal wages is cheaper transport into and from the country, less profit to those who sit down and do nothing, and a resident population.

In the summer of 1908 much was heard concerning 'muck discoveries', that is, a discovery based on the finding of particles of gold in the black vegetable matter under the moss and

A CLEAN-UP ON CLEARY CREEK.

above the gravel. This 'muck' covers the country like a dirty blanket. According to law, a valid discovery necessitates the finding of mineral in place. In placer mining *the* mineral is native gold, therefore the dispute involves the question whether the gold particles in the muck are part of the deposit which the miner is seeking, or whether it is an accidental occurrence in no way related to the actual gravel deposit underneath. The prospectors ridicule a 'muck discovery', claiming that the gold in the overburden is accidental and no more significant than if found in the bark of a neighboring tree; indeed, gold has been found in clear ice. The lawyers interpret the evidence otherwise, claiming that any gold found on the claim is adequate to establish a proper title. Incidentally, it may be mentioned that many shrewd prospectors when they locate a claim attach a lawyer's name, as well as their own, to the location notice, arguing that it is cheaper to have a lawyer for a partner than to hire him later. The cry of the average unlettered miner is like that of the French at the outset of a revolution: "*A bas les sacrés avocats.*" Surprising is it not; it is not.

CHAPTER XXVI.

ARCTIC AGRICULTURE.

On August 10 we left Fairbanks on the *Tanana* and went down the river to a point 80 miles above Fort Gibbon; disembarking, we boarded a small steam-launch that took us 7 miles eastward up a slough to Manley's Hot Springs. This is a notable resort. A hotel, bath-house, and farm have been established by Frank G. Manley, a successful miner, who co-operates in the management of the property with J. F. Karshner. This old prospector, formerly a farmer in Kansas, discovered a hot spring and ascertained that the ground near it was warmed enough not to freeze even in winter. He took up a homestead of 320 acres, and in the spring of 1903 made a clearing for a garden; then, his first effort being surprisingly successful, he commenced systematic cultivation. Later Manley established a trading post on the homestead, building a hotel and bath-house in 1907. At this spot there stood a grove of evergreen poplar, known as Balm of Gilead, and from them were sawed the logs for the erection of the roomy hotel building—a sightly structure, as the accompanying photograph shows. The adjoining bath-house contains two small plunges and a large swimming pool. The water has a temperature ranging from 115°F. in the small baths to 100° in the large one. The springs are about a hundred yards from the hotel, and from them the water issuing at 135° is piped as required. On the adjacent hill-slope the ground is warmed by the thermal springs so as to afford abnormal conditions highly favorable to agriculture. It is a natural hot-bed. Snow falling on the warm ground thaws rapidly so that the surface is rarely white for more than a couple of days. At the most the surface-frost penetrates only to a depth of one inch. On the edge of the warm area, young parsnips have been dug in March from under two feet of snow. The water issuing from one spring will fill an 8-inch pipe and another spring higher up will fill a 3-inch pipe. A pent-house has been made by cutting into the side of the hill, where the ground is heated sufficiently to allow poultry to live in comfort throughout the winter. Mr. Manley has 650 hens and 50 ducks, 8 cows and a bull, 70 pigs, and 25 horses. To appreciate what this signifies it must be remembered that this farm is in latitude 65° north. Altogether, 62 acres are under cultivation; of these, 32 are devoted to potatoes, which retail at 12½ cents per pound at Fairbanks. Usually the yield is 7 to 8 tons per acre, and a ton of potatoes in this part of the world is worth $250, be it noted. In 1908 the lack of rain injured the crop, so that the yield was only 3 tons per acre. Five acres are given to turnips, lettuce, cabbage, and carrots. Fodder is obtained from 14 acres of oats, barley, and wheat. Near the springs the ground is covered with wild peas and bracken, for the moss does not thrive on the warm land. The soil is a sandy loam of light chocolate color, derived from the disintegration of granite.

As indicating agricultural possibilities I cite the experience of J. C. Riley, of Tolovana, who planted half a crate of potatoes in 1908 and obtained 11 crates therefrom; in 1907 he planted an equal amount and the ground yielded him 16 crates of marketable potatoes, worth 12 to 15 cents per pound or $250 to $300 per ton. That is the price for which they sold at Fairbanks. Last season, a woman living at Gilmore, on Gold Stream, cleared one acre and sold $3000 worth of potatoes from it. Apparently there is no excuse for failure to cultivate locally at a profit, but the fact is the population is migratory and extravagant, and also intimidated by the big trading companies, which may at short notice spoil the local market for any commodity. Moreover, the indigenous crop is uncertain, making it necessary for the provident to order their staple food-supplies from the 'outside' two or three months before the opening of the season.

Hay is worth $125 per ton. Native red top sells for $80 to $100 per ton. Other necessaries, such as cabbage, turnips, dairy products, eggs, can be produced locally at a large profit, if grown at all, for the freight from Seattle and San Francisco represents a margin big enough to make the indigenous article highly remunerative.

These successful attempts at cultivation prompt an enquiry into the possibilities of agriculture in Alaska. For light on the subject I went to C. C. Georgeson, Special Agent in charge of the Alaska Experiment Stations, and to him I owe most of

MANLEY'S HOTEL, HOT SPRINGS.

the information that follows. Broadly speaking, any kind of hardy vegetable can be grown even as far north as the Arctic Circle, 66°33′ north latitude. During the summer season, which varies from 3 months at Rampart to 6 months at Sitka, fodder is obtainable from a luxuriant growth of native grasses, timothy, and oats. It is true, on the coast there is trouble in making hay on account of rains at the gathering period, especially at Sitka, where the annual precipitation is 8 feet. 'Inside,' within the vast interior of the country, the light precipitation may necessitate the aid of irrigation; thus in the Copper River valley the precipitation is only 9½ inches annually, and the lack of water is a drawback. On the other hand, the ground of interior Alaska is frozen to a depth of 200 feet, and it is found that the subsoil ice actually assists cultivation in summer because as the thaw proceeds the moisture rises by capillarity to the roots of plants. The most northern experiment station, at Rampart, in Lat. 65°40′, was begun in 1900. On a tract of 6½ acres, it has been found possible to mature barley and oats from year to year. In addition, winter rye, winter wheat, spring rye, spring wheat, and buckwheat have come to maturity in three years out of four. There are now 16 acres under cultivation at this station. At Rampart, and elsewhere in the Yukon, the settlers have grown vegetables with unqualified success. Potatoes always do well, and they are doing better as potatoes grown in Alaska are used for seeding. It is a fact that this staple tuber has been grown 60 miles north of the Arctic Circle; that is as far as any pioneer gardener has yet ventured. Cabbages also, with cauliflower, peas, turnips, radishes, lettuce, carrots, parsnips, parsley, beets, onions, squash, and rhubarb, all flourish during the short warm summer. In the more favored regions, such as the Tanana, it has been found that beans, celery, cucumbers, and salsify will grow well; even cucumbers, musk melons, squash, and water melons have matured outdoors at the Hot Springs farm. Forage crops for livestock can be cultivated successfully. Timothy springs up as a volunteer crop along every trail where hay has been carried. In many localities the nutritious grasses native to the country cover large areas with lush growth, affording rich pasturage. At Rampart the horses employed at the experiment station are fed on native hay in winter. Emphasis must be placed on the fact that the seeds from which crops have been obtained heretofore in Alaska are of southern origin, that is, from a latitude fully 20° farther south; in consequence, the plants have not been adapted to the climatic conditions of Alaska, and it is reasonable to expect better results from northern seed. For this purpose the experiment stations are well fitted, and it is likely that they will do useful service in

VEGETABLES GROWN AT LATITUDE 64°51′ NORTH.

the development of varieties suited to the new environment. This beneficent work has been well started, but it is hampered by the need of funds. The first appropriation voted by Congress was only $10,000, and although the sum was increased to $15,000, it has barely sufficed to start and equip the five experiment stations at Rampart, Fairbanks, Copper River, Kodiak, and Sitka. The Kodiak station, which is newly established, is devoted to the breeding of livestock, and to this end has been provided with well bred Galloway cattle. These run free all winter and are fed only when the snow covers the ground, and then on native hay. Another station has been in operation at Kenai, on Cook Inlet, for nine years. It was demonstrated that grain would not mature in that climate; but on the other hand, that live stock could be kept successfully. A dairy was operated on the station and first-class butter and cheese is made from the milk of cows fed exclusively on native grown fodder. The station is now closed and the stock transferred to Kodiak station. Sitka station, which is also headquarters for the agricultural investigations, is devoted to experiments in horticulture, while the interior stations above mentioned are reserved chiefly for the culture and breeding of grains.

Back from the coast the soils are commonly of a light loamy character in the valleys, becoming gravelly and thin on the higher ground. The river bottoms also afford extensive areas containing a mixture of silt and fine sand, which is exceedingly fertile. These are overgrown with willows stretching back to the universal wall of the spruce forests. The soils of the interior are entirely free from 'sourness', which affects much of the coast land. Thus no lime is needed.

It is estimated that the interior of Alaska contains 90,000 square miles available for pasturage and agriculture; this area is nearly equal to the two States of New York and Pennsylvania. In 1900 New York had a population of 152 persons per square mile and Pennsylvania had 140. Of the whole of Alaska only one sixth is deemed fit for cultivation, and it is reasonable to hope that this favored portion of the country will eventually support 25 persons per square mile, this being the average density of population in the United States (exclusive of Hawaii and Alaska) in 1900, even including the deserts, mountains, and forests now uninhabited. If interior Alaska could yield products sufficient to support 25 persons per square mile, the population could rise safely to 2,250,000. Now it is about 20,000.

For a parallel we can go to Finland, which is bounded on the south by latitude 60° and on the north by latitude 70°. Finland has an area of 148,000 square miles, or about one quarter of Alaska. Of Finland, one third is lake and marsh

MANLEY'S HOT SPRINGS.

land, another third is covered with forest, leaving only about 50,000 square miles available for agriculture. Yet this country in northern Europe supports a population of 3,000,000, or 60 persons per square mile utilized for cultivation. Why should Alaska be less productive? Temperature is a controlling factor. From this standpoint, it is interesting to note that at Helsingfors, the southernmost point in Finland, the average annual temperature is 38.7°F.; at Sitka, it is 43.3°. In northern Finland the average annual temperature is 27.05°F.; at Rampart, it is 27.50°. Finland exports butter, beef, bacon, and even grain, chiefly oats. There is hope for Alaskan agriculture. If it has made but little progress as yet, the failure is due largely to the greater attractiveness of the Canadian Northwest, and to the great expense of transport to, and in, Alaska. So far the farming in Alaska has been done by disappointed prospectors and diggers who seize the opportunity of making a little money by growing garden truck for the mining camps.

CHAPTER XXVII.

ON THE LOWER YUKON.

We spent a week at Hot Springs, five days more than planned, for the movements of the river steamers are irregular and the agents of the Northern Commercial Co. inform travelers that the receipts from passenger traffic are of no consequence compared to the freightage of supplies. Yet, on the *Sarah* there were 108 first-class passengers, 67 second-class, and 100 deck passengers. The larger steamers, or 'packets,' carry an average of 100 passengers at $80 and push two barges loaded with 1000 to 1500 tons, the freightage on which is $75 per ton, from Seattle or San Francisco to Dawson or Fairbanks, and of the $75 only $10 is the cost of ocean transport to St. Michael. Thus receipts of $75,000 to $90,000 per trip are indicated. As against this, the expenses of a 'packet' for the entire summer season are $70,000, or $800 per day. Apparently there is a margin for big profits and no adequate excuse for the high cost of transport, which is today one of the principal obstacles to the settlement of the interior portion of Alaska. The fact is the N. A. T. and N. C. companies, as the two chief trading concerns are popularly called, control the sale and transport of the necessaries of life throughout the interior; by combining and by driving out competitors they are able to prevent interference with their control. At Nome, 2700 miles from Seattle or San Francisco, prices are but little higher than in the two commercial centres of the Coast, but at Fairbanks and Dawson, owing to control of the river traffic, everything is preposterously high in price. From Nome to Fairbanks the distance is only 1150 miles or less than half that to San Francisco, and while allowance must be made for steaming against the river current, the main reason for higher freight rates is the fact of a monopoly shared by the two trading companies. St. Michael is a military reservation and no warehouses can be maintained there without permission of the War Department; in consequence, the right is denied to any but the two trading companies mentioned. As an example of the spirit shown by these concerns, I cite the following true story: In 1897 L. R. Fulda, going up the Yukon to start work for the Alaska Exploration Co. arrived at St. Michael just as the last boat for the season was going to Dawson. It was a N. A. T. steamer and J. J. Healy of that company instructed the captain not to give Fulda a passage, for he guessed his purpose. Thereupon Fulda changed clothes with a longshoreman and applied to the mate for work; being athletic and willing, he did well. Finally he signed ship's articles and shipped as a deck-hand. The steamer started. Healy was on board; recognizing Fulda, he told the captain to put Fulda ashore at the next landing; the captain gave the order to the mate, but the latter objected because Fulda was a good worker and could carry more wood than any other deck-hand, besides he had signed articles and thus possessed a valid contract. In this way Fulda reached Dawson. Within twelve months he had charge of a big river business, including 7 or 8 steamboats and 15 barges; he controlled transportation and trade amounting to $1,000,000 per annum. "He got there." The story exemplifies the persistence of an energetic man and the lawless spirit of a dominant corporation. It is a legacy from the days of the Russian American Co. and the Hudson's Bay Co.; it is an anachronism in a civilized democracy. The administration of the N. A. T. and N. C. companies needs to be investigated and disciplined, not without recognition of the skill and resource exhibited by their agents in the early development of the region, but with an eye upon their relations to United States senators.

While the above was being written we were on the *Schwatka* moving swiftly down the Tanana toward Fort Gibbon, where connection is to be made with the packet *Sarah*. The little *Schwatka* was crowded. Many 'sporting ladies' were aboard and they were noisy as usual. We slept in bed linen smelling of stale food, for the sheets had been rescued from among the soiled table-cloths with which they had been heaped in preparation for the 'wash.' This is one of the imperfections of travel on the Yukon. While unwilling to make a futile protest and observant of the kind of passenger traffic for which the boats are mainly intended, I could not but regret that the intrepid geographer Frederick Schwatka should be commemorated on the noble river of his exploration by nothing better than a miserable little steamer with a stern-paddle and dirty linen. But that reminds me that I ran across a more pleasant re-

STEAMERS AT FORT GIBBON.

minder of Schwatka: in the pilot-house of the *Lavelle Young* there was a blue print from Schwatka's original map of the Yukon. Capt. Boerner informed me that he found it useful for reference. Although a rough bit of surveying, the map is wonderfully correct; it shows Schwatka's camps and his daily runs. However, we were soon transferred to another boat. Arriving that night at Tanana, we were awakened next morning by the cheerful bugles of the fort. A stroll through the town or, more accurately, along the single street facing the river, yielded many interesting sights, although we confess to having missed "the streets of tropic bloom" described by a recent writer, just as we had made no acquaintance either with the "metropolitan style" of Fairbanks or "the luxurious steamers" that ply thither. Travelers' tales are often made sweet with rhetorical confectionery. Alaska does not need such literary treatment, for there is enough of interest without exaggerating.

At the Army post we inspected the dog-kennels, occupied by wolfish malamutes, muscular huskies, and 'outside' dogs of all sorts. The malamute is the Eskimo dog and is named from a tribe on the lower Yukon, while the 'husky' came from the Mackenzie river. They are of ordinary size, but stocky in shape, sturdy, and muscular, well protected from the cold by thick woolly hair. Summer is hard on these long-haired beasts, they lie on bare ground to receive the cold of the underlying ice or else find a place under the bank where the overhanging moss gives shelter from the sun. On warm days the dogs lie panting in the shade and only at evening do they bestir themselves, or when a steamboat arrives. As soon as the whistle is sounded they make for the landing, eager for the scraps of food thrown ashore by the stewards and cooks. The dogs do not howl at the regular town whistles, but the salute of an incoming steamboat or any unfamiliar spectacle, such as a fire, will cause them to set up a melancholy chorus. At intervals, and unaccountably, the quiet night air of a town, like Dawson, will be rent by a sudden outburst of howls from the uneasy malamutes and huskies, who will subside into slumberous silence after two or three minutes. The reason is no more obvious than the sudden stampeding of cattle or the crowing of cocks in the middle of the night, although the nocturnal concerts of the native dogs have some resemblance to the weird howling of the coyotes at dawn.

The malamute knows not how to bark, he can only howl. The following story is apropos: At Nome in the spring of 1899 a setter had a litter of five pups, the father of whom was obviously a malamute. The mother dog was proud of her family and used to bring them into the Ames Mercantile Co's store. One day, when the pups were big enough to run about, she discovered that they howled, but could not bark; this evidently disgusted her. Soon afterward she was seen with her five pups in line on the beach teaching them to bark: she barked and then looked at them as if giving them a kindergarten lesson. Many persons saw this performance and noted the progress of the experiment. Finally, one pup did actually bark, much to the delight of the mother, as she indicated by wagging her tail and jumping about. Soon all the pups learned to bark like civilized dogs.

But the malamute is a savage and is devoid of those instincts of faithfulness that make the dog a friend of man, although Jack London can spin fanciful yarns about him. Here is a true story, to offset London's tales. An 'outside' dog had

MALAMUTES IN CHORUS.

seven pups, of malamute breed; one day the mother got her paw into a hole between the logs and howled in pain; her pups attacked her, killed her, and tore her to pieces before an observer could drive them off. In summer the dogs become bad tempered, fight among themselves, and attack children. While at Nome, I saw a crowd gathered around a dog held by a policeman. Bystanders informed me that the dog had snapped at an Eskimo child, and when the testimony was clear the policeman hauled the dog by the scruff of the neck to the beach and shot him. The malamute's only motto is "Woe to the vanquished;" if one of his brothers goes down in a fight, or accident, the others all jump on him at once. This is what makes it dangerous for children to play with them, for they will jump on a child that stumbles. In the summer of 1907 a boy eleven years old was carrying a dinner pail to his father, who was at work at a mine on the tundra a short distance from the town of Nome. The boy was accompanied by four dogs; he happened to stumble, and fell; thereupon the dogs pounced upon him and rent him. Again, the inference is that the malamute is an utterly unredeemed savage; he is to the dog tribe what his master, the Eskimo, is to the human species.

Enough of dogs; let us proceed along the river bank and see the camp of the prospectors on the outskirts of the town. A dozen tents are occupied by men awaiting a chance to go up the Innoko, the Kantishna, or Sullivan creek—these being the latest 'stampedes.' Looking at these adventurous men, not particularly robust in appearance nor particularly cheerful, the idea comes that their lot is not to be envied, that their life is an exile and their existence barren of comfort; and yet they would not thank you for sympathy. They do not need it. To these hardy explorers the life of the nomad is attractive and the lure of the 'creek' an excitement that is constantly renewed; the reward of gold is sufficient to them, for with it they buy all the creature comforts for which they care—cigars, whisky, women. A rich gravel claim going $2 to $4 per square foot 30 feet deep in frozen ground is as good as a potato patch ready to be gathered and marketed among hungry men. "*Chacun à son goût et le marchand vend tout.*" The rush for gold on the creeks is no more ignoble than the similar scramble every day on State Street or Wall Street; the pick is as fine an instrument as the ticker, and the steam-point is as useful as the tape.

But these thoughts drop into oblivion as the day wanes and the panoply of the sunset is spread athwart the gateway of the West. The blue vault of the upper sky merges through emerald tints into the golden splendor of clouds radiant with the glint of the sinking sun. Flat masses of blue-gray mist float silhouetted against the fading brightness, which illumines the rolling contours of the wilderness, now purple in the twilight. The great river sweeps around a headland; far away, faint but clear, is a mountain range parting the beauty of the earth and sky. Soon the last rays of the sun are quenched and a dreaming radiance robes the vast Northland in light that never before was on land or sea.

Next day we waited for the *Sarah*, but she was delayed. Late at night she came in sight, and on arrival we were transferred, expecting more comfortable quarters, for was not this one of the "big packets," of which the agent at Dawson had spoken rapturously. On arrival the *Sarah's* whistle caused all the dogs in town to set up a melancholy howl, and on recalling my journey on the *Sarah*, her accommodations, her service, and her food, I fain would imitate those malamutes as the only way of expressing my feelings.

However, the *Sarah* finally got away on the day after her arrival at Tanana and the last stage of our journey on the Yukon was begun. The river swept through a flood plain bounded by wooded hills. On the banks the effect of the ice is seen in the removal of evergreens, such as spruce, causing a fringe of willow to stand between the river and the forest. Former sloughs are indicated by a young growth of willows or a glade of waving grass. Clear streams enter the muddy main river and force a contrast.

At Nulato we touch the page of history, for this is an old trading post. The present settlement consists of a telegraph station and two stores, a number of Indian shacks, the residences of a U. S. Government doctor and a Russian priest, with a log church, ornamented in the tawdry style calculated to impress an Eskimo. The buildings are sufficiently weather-beaten to appear ancient, but they are all recent. The old settlement was a quarter of a mile below the present village, the site of it being now marked by a clearing covered with brown grass in which will be found 15 holes, indicating the 'dug-outs' or *igloos* of the former Russian camp. All vestiges of the old Nulato are gone save four graves, in which lie three Russian traders and an English officer.

Originally known as Fort Derabin, from its founder, a Russian, it was a post of the Russian American Company in 1841, having been rebuilt on the site of a trading station established by Malakhoff, a Russian creole, in 1838. Mention has been made of the ascent of the Yukon by Lieut. Zagoskin, of the Russian navy, who came to Nulato in 1842. For ten years this lonesome little outpost served as a mart for the furs brought in from the surrounding country by the Indians. In 1851 Lieut. Barnard, of the English ship *Enterprise*, arrived in search of Sir John Franklin. Making a remark that was misunderstood, Barnard angered one of the Indian chiefs. A sudden attack was made on the post and all the inmates except one young man and two or three children were massacred. Barnard and Derabin lie in their abandoned graves, within a stone's throw of the stockaded fort that in 1859 replaced the former log-houses. The natives who died in the massacre were buried in the holes where the Indian houses used to stand. The graves and the holes are all that survive, and over them the rank grass has spread an obliterating mantle. Abandoned houses do not last long in Alaska; they are too valuable for firewood.

Just above the present village of Nulato a cluster of multicolored Indian graves decorates the crest of a ridge overlooking the river. In their red, white, blue, and green these look like doll's houses, inside of them food is placed, and alongside stand crosses, brilliantly tinted also. In their gaudy atrocity they serve as landmarks to those who travel on the river. A mile below another group of graves and caches marks the site of the reindeer village belonging to the U. S. Government, but now in charge of the mission. In summer the reindeer are herded at Holy Cross, but in winter they are brought to Nulato under the charge of Dr. Norton. There are 10,000 reindeer in Alaska. About 28 miles up the river is the mouth of the Koyukuk and a village of the same name is near-by. Good news of the gold diggings has been received, but the prospectors are said to be short of provisions.

Those were long days on the lordly Yukon; I have a vivid memory of the minor happenings that were repeated continually during the voyage. The regular deep breathing of the engine as the steam issued from the exhaust; at intervals the rattle of the cable of the steering gear, when the pilot turned the boat in the sinuous channel; at the end of every half-hour an explosive rush of steam as the mud was washed out of the boiler; the desultory talk of passengers; the fragrance of a briar pipe; a low shore and a silent land; the scrubby forest of spruce and the distant ridge of hills; a high bank undermined by the current, with trees lying prone on the river's edge; a white tent and a fish-wheel; the splendid splash of pink flowers in the middle distance and the dark cloud of a forest fire far away; sun and air, vivid and vivifying; rapid and continuous movement into a vast wilderness; a feeling of mental and physical alertness, with a preparedness for any-

NULATO.

thing that might happen; and through it all the strong regular respiration of the tireless engines that were conquering the successive miles of travel.

Approaching Kaltag we passed two Indians, a man and a woman in a boat; they were 'poling' and their progress was further aided by two dogs that towed the boat laboriously against the stream. A little farther we met another party traveling in the same manner. On the left bank a group of tents marked a fishing camp, as we could also tell from the red patches of salmon hung on frames preparatory to being smoked.

Then to the right on a high bank we saw three or four buildings, one of which was the station of Kaltag, an important point in the telegraph service. The *Sarah* blew her whistle, and before the echoes had died a long-drawn melancholy howl, as of lost souls in the pit of torment, came from the shore. It was the malamute dogs.

At Kaltag the Yukon makes a big bend southward, so that while it is 570 miles by water to St. Michael it is only 90 miles across country to tide-water. It has been suggested that a railroad across this low portage would effect the crossing in one day easily, and thus save two days, at least, for the boats are often delayed by storms and by going aground on the bar at the mouth of the Yukon. The telegraph line runs straight from Kaltag to Unalaklik, on Norton Sound, a distance of only 90 miles. The stakes set in 1867 by the expedition of the Western Union Telegraph Company are still visible and at the seaward end of the portage, 4 miles north of Unalaklik, the ruins of a station survive, together with a few poles and some wire.

Anvik, the trading post for the Innoko, was passed in the night, so we missed meeting Max Simel, a squaw-man and a notable trader in these parts. His chief rival was Lon Cooper, who bought fish from the Indians on the basis of 30 for a dollar, the regular price being 20 for a dollar. Simel held back until the end of the season and then offered to buy at 20 fish for a dollar, much to the discomfiture of Cooper, for, of course, the Indians responded quickly. All payments were made not in coin but in staples, chiefly tea. So, Simel, in paying, measured a half-pound of tea, instead of a pound, whereupon the Indian demurred to the smallness of the package; but Stimel explained that it was a 'fish-dollar.' He worked the same game when selling reindeer skins: Cooper took two martens for one reindeer, while Simel asked three martens for his reindeer-dollar. Thus a new application of the trade dollar was successfully made and Jerusalem was justified of her children.

At Holy Cross a Russian mission lingers and the tawdry church and chapel of the Greek church dominate a neat little settlement, the cleanliness of which is in strong contrast to the

HAULING FREIGHT OVER THE ICE OF BERING SEA.

Indian fishing villages. A black-robed half-breed priest, with long black hair, a black smock, and a bowler or derby hat, came aboard—a curious anachronism and a depressing object.

Below Holy Cross the Yukon emerges into the flats of the delta, which reaches inland for 60 miles. Here navigation is impeded by silting of the channels, for the river has many outlets meandering through marshes and islands, so that its flood is abated and spread over a large area, causing shallow water. During our journey we had seen how the banks are undermined and the forest swept into the river each spring, so that it was easy to understand how big a mass of débris is brought down each year for deposition on the sea coast. For five miles from shore soundings show only 4 feet, the cause of which is seen in the waters discolored by the sediment that settles far out to sea.

Even in the delta the scenery, though tame, has a quiet charm, heightened by the prospect of ending the journey. The yellow marshes, the vividly green brush, the flocks of geese in long procession, and the blue bourne of hills on the eastern horizon give the picture a touch of dignity and a feeling of spaciousness. Soon we approached the sea. On the right the marshes of the delta terminated in the bold headland of Cape Romanoff, while on the left the shallows of the estuary merged far away into the blue of the open sea. Turning northward we reached St. Michael, four days from Tanana.

CHAPTER XXVIII.

ST. MICHAEL AND NOME.

St. Michael is known to every traveler in the North; it is a name to conjure with, for it evokes both curses and praises. Of all the ports of call on the main lines of travel about the world there is no place viewed so subjectively as St. Michael.

ST. MICHAEL.

The personal point of view colors the impression of everyone who ever landed on that lonely island in Bering Sea; for example, Mr. Jeremiah Lynch found "a poor hotel," "innumerable dogs, guarded by a few squalid Indians." He had to stay there 12 days: "It was a dreary detention." On the other hand, Mrs. Ella Higginson found "an excellent hotel at St. Michael," also "beautiful ivory carvings" and some gorgeous sunsets. Her mood was most sympathetic: "The tundra is rolling, with numerous pools that flame like brass at sunset," and so forth. Finally, the lady waxes rhapsodical: "In all the world there cannot be another spot so noble in which to lie down and rest when life's fevers and life's passions all are past." This author is in evident agreement with the steamship companies, which have a way of compelling travelers to spend a few days between boats on a spot the beauties of which are not visible to the casual observer. St. Michael is a place of "dreary detention," as Mr. Lynch truly says; it looks like a penal settlement, and while it may be an excellent spot when "life's passions" are over, it is a most undesirable place of sojourn as long as anyone can bribe or hire a vessel to carry him elsewhere. The hotel is execrable, the town is decrepit, the barracks are hideous; the derelicts rotting in the harbor, the filthy Indians slouching on the shore, the soggy morass in three directions and the gray sea in the offing, all combine to make a picture that has been known to excite a "fitful fever" of great violence and unexampled eloquence against things in general.

However, I ought not to say too many unkind things about St. Michael, for I escaped detention, spending only six hours there. During that interval I saw some of the relics of the Russian occupation. A fort was planted here by Michael Tebenkoff in 1833, the block-house being built of drift-logs brought down by the Yukon. This block-house stands on the shore; inside of it there remain six toy-like cannon, weighing about 100 pounds each, on wooden carriages with wooden wheels; and the gunports indicate how they were used. A Greek church survives; the ministrations are in the hands of two Russian priests, the congregation consisting of a dozen Eskimo and a stray tourist. At the back of the hotel is a small fenced enclosure from which the body of a Russian bishop was taken to be interred in front of the church, the body and coffin being found preserved in solid ice. In digging, the wood of the coffin was broken, and thus by accident the body was exposed to view, proving to be

LANDING PASSENGERS AT NOME.

in perfect preservation and recognizable as one of the earliest Russian ecclesiastics.

The shore is fringed with blocks of black basalt and on the horizon several conical hills suggest volcanic vents. In the topography the lava flows, covered by tundra, are manifest. Geologically, St. Michael is recent, for it was raised above the sea not so very long ago, the natives even alleging that it has been submerged within the memory of their forefathers.

We were fortunate in catching the *Victoria*, of the Alaska Steamship line, reaching Nome, 106 miles from St. Michael, in 9 hours. To arrive at Nome is not as simple as it sounds. It was night; while the *Victoria* was yet churning the cold waters of Bering Sea we saw the electric lights of the town, a flashing coronet on the cold brow of the North. After nearly a week's journeying down the sullen Yukon and through the heart of the inhuman wilderness it was a pleasant shock to see the evidence of modern industry and to be made aware of this brave little community of adventurers so far from civilization.

After an hour's wait, and signaling with the shore, two miles to leeward, we saw a lighter towed by a tug approach within the field of the *Victoria's* searchlight. Disembarking, we descended onto the lighter, which was swinging in a gentle swell. All being aboard, the tug towed the crowded lighter, but the latter having no steering, she swung first to one side and then to the other as the waves willed. When about 150 yards from the shore and within sound of the surf, we came close to a tall steel tower standing in the shallow sea; it proved to be the terminal stations of an aerial tramway, used for transporting freight, and it also served as the anchorage for an endless hemp rope that ran to the beach. The men in charge of the lighter grappled for this rope, but, after several attempts, failed. While this performance was under way the lighter was drifting ominously near the line of the breakers, so that the tug had to be recalled to pull the lighter back over the line of the cable. Then finally the cable was caught, the attachment was made secure, and the lighter was pulled rapidly (by the endless rope) to the landing stage; a gang-plank was dropped expeditiously, while the lighter rose and fell to the motion of

FRONT STREET, NOME.

the sea, and without more ado we scrambled up the gang-plank to the high wharf. Thus we reached Nome, the most northern of the great gold mining camps of the modern world.

Next morning (August 26) a ride on horseback over the tundra yielded first impressions of the environments of Nome. Leaving the planked streets of the town we followed the wagon-road built by the Government across the tundra to the mines on the Third Beach. This road is neither macadam nor corduroy, it has simply been drained and graveled, with a result not wholly satisfactory, leaving a broad black streak stretching northward to the hills on the near horizon. Looking across the soft contours of this coastal plain, carpeted with moss, the tundra is seen to stretch to a number of ridges indented by valleys. Between Anvil mountain and Newton peak, both of which rise slightly above 1000 feet, is Dry creek. The seaward slopes are broken by gray outcrops of limestone, while on the crests of the hills fantastic shapes have been weathered out of schist, one of these being the celebrated Anvil rock, which gave a name to the wonderfully rich valley in which the pioneers found fortunes.

On leaving the town the road runs between the shallow depressions along which Dry and Bourbon creeks find their way to the sea; by reason of a natural concentrating process these small valleys contain deposits of gold-bearing gravel rich enough to be mined by various methods. A large dredge, evidently new, lies with broken back in its own pond in Bourbon creek, telling of a sad fiasco, while close-by two or three ruined contrivances of queer shape indicate ill-advised efforts at early dredging. On the tundra in this vicinity an idle Keystone drill suggests the service done to mining exploration by this useful device, whereby a columnar sample of the gravel deposit is obtained before actual mining begins. Riding on, the next object to arrest attention is a cluster of stakes—six of them—indicating a conflict of ownership, for while four stakes would be required at the corner where two bench claims and two creek claims intersect, the two extra posts mean that someone had been 'jumping' the location, with an inevitable sequel of litigation, such as has embittered the whole existence of this frontier community.

Leaving the main road we follow a trail along Dry creek, where mining is in progress. A group of men are shoveling the gravel into sluice-boxes, the water for washing being conducted from a ditch through a canvas hose, whose white serpentine length quivers with life as the water courses through it. This method is simple and flexible; it was used by the pioneers in California. The hose is 14 inches in diameter and is made of 12 to 14 ounce canvas, sewed with three seams.

DISABLED DREDGE ON BOURBON CREEK, NOME.

Where the men are mining, the gravel has thawed naturally; this is indicated by clusters of scrubby willow, little bushes only three or four feet high, but a valuable sign to the miner, who knows that wherever on the tundra he sees the dwarf willows he can be assured of soft ground all the way to bed-rock.

Surmounting the rise above Dry creek the trail crosses the line of a railroad, a branch of the Seward Peninsula Railway. This narrow-gauge system was originally built, to Anvil creek, by the Wild Goose Mining & Trading Co. at the time when Charles D. Lane was in command; it is 80 miles long, with the inland terminus at Lanes Landing in the Kougarok district. While serving a useful purpose, this is a railway of crude construction, without rock ballast and without grading, laid simply on the rolling tundra. The rails lie on ties and the ties on 'stringers' of three-inch plank placed lengthwise. On slopes only one stringer is used, namely, on the lower side. The line is ballasted with moss and follows the easiest available contours, up and down and round about. An attempt was made to strip the moss and lay the track on the gravel, but this plan proved a failure owing to the melting of the exposed surface causing the impermanent way to sink out of sight. I did not travel on this railroad, although in receipt of a card that would have made a journey inexpensive. Nor was this declination unwarranted, as was proved by a sad fatality a few days later. On September 15 a train going to Lanes Landing was derailed without any apparent cause, and when the cars were overturned several persons were seriously injured; among these was Cabel Whitehead, who two or three days earlier had resigned as president of the company controlling the railway. Whitehead was a chemist and banker, a leader in the development of the Seward Peninsula and a notable figure in Alaskan history. He had invited me to go with him on a fishing and hunting expedition, but another engagement prevented acceptance of this courtesy. When the train 'jumped the track,' he was sitting on a flat car, on which lumber was loaded, and as the car turned over he was thrown into a water hole in the tundra with some of the lumber on top of him, so that fully two minutes elapsed before he could be extricated. His lungs had become filled with the sandy cold water, violent pneumonia supervened, and he died two days later. The funeral took place on the day when the *Northwestern* sailed, and as we lay off shore on September 8, a lighter was towed alongside, and a coffin covered with the national flag was hoisted aboard, with no further ceremony than the doffing of every hat. He was a useful man, a scholar, and a gentleman.

CHAPTER XXIX.

NOME AND THE ESKIMO.

Nome is attractive, a haven alike for the storm-tossed sailor on Bering Sea and the leg-weary traveler over the interminable tundra. It is a clean little town inhabited by a cheerful com-

NOME.

munity of hardy people. Stretching along the edge of the coastal plain, Nome is on the fringe of things in general, but it has a strong grip on life and happiness. Being on the tundra, the streets are planked, otherwise they would be mere canals of bottomless mud. Owing to the mode of construction and the scarcity of vehicular traffic, the streets are quiet and clean.

The main thoroughfare is Front street, which is narrow and crooked, giving picturesque effects. For a mile it extends between the wooden dwellings and shacks dedicated to many and varied uses, from banks to bagnios, from stores to saloons, from fish to furs.

Nome has a population of 4500 in summer, when business is most active; in winter the number shrinks to 3000. In 1908 the town polled 1500 voters. Nome boasts a larger proportion of women and children than any other settlement in the Far North, and despite the yearly exodus there is a steady increase of those who are willing to remain through the long winter. Most of the engineers and other professional men connected with mining operations go 'outside' just before the close of navigation, as fixed by the southward movement of the ice-pack. In 1908 the "last boat" left on October 23. From then until June all communication is overland by dog-team to St. Michael and thence by way of Fairbanks to Valdez, where steamship connection is made with Seattle. Early in June the "first boat" arrives amid great excitement. In the effort to bring the first stock of supplies to the hibernating folk at Nome, the steamers from San Francisco and Seattle have been known to brave the ice-pack, with results disastrous to everyone concerned. In 1903 the *Portland* and the *Jennie* were caught in the ice and were carried through Bering strait into the Arctic until the ice-drift released them. In 1908 the *Yucatan* was 17 days overdue and the *Victoria* arrived with a hole in her bow, while the ice-pack gripped the *Ohio* so securely that she was 33 days overdue. Whether the ice was wholly to blame, or an over-cautious captain of the *Ohio*, is a subject that will provoke violent discussion at Nome even to this day.

When called into being as a place of landing at the mouth of Snake river, in 1899, Nome consisted of a few tents, to which log-cabins were added when the pioneers had collected the drift-wood on the shore. In the first winter no other fuel was available except the drift-wood brought by the Yukon into Bering Sea and strewn by the waves along the low coast. It was a providential gift. Even with the help of wood fires that was a dreary winter for the handful of men who camped on the

ESKIMO WOMAN AND CHILD.

edge of the coastal plain. They tell me that there was much fog, causing men to lose their way on the trackless waste; among these was the present City Attorney, who was lost for four days. Once away from the shore there was nothing to serve as a landmark and not many were wise enough to use a compass. The drinking of tundra water and the general unsanitary conditions led to an outbreak of typhoid, from which 100 died at Nome, while others died on the steamers going home. This unnerved many of the newcomers, but it had a good result in stimulating the construction of a pipe-line to the Moonlight springs, whence excellent water is brought four miles to the town.

Those gold-seekers who went 'out' in the fall of 1899 and landed at San Francisco and Seattle announced the richness of the beach and the creeks of Nome. As nearly $3,000,000 worth of gold testified to the truth of their stories, the interest excited was keen enough to cause a big stampede, like the one to the Klondike two years earlier. But it differed from that 'rush' in the fact that access to the new diggings was wholly by sea, and without privation, so that the crowd of men who landed on the beach of Nome in June 1900 were even less fitted for rough work than those who tramped over the White and Chilkoot passes.

Every calling in life was represented, especially the shiftless and visionary who hoped to make a short cut to wealth. They paid from $125 to $300 to come to Nome and they paid from $100 to $125 to return to 'the States,' as most of them did after a rapid disillusionment. Steerage passengers paid $60 to $75. The steamers were terribly crowded, 1100 to 1200 men were put on boats of 2000 to 2500 tons, and every old hulk on the Pacific coast was requisitioned for this lucrative service. There was the inevitable 'graft' that disgraces every episode of this kind. Empty cabins on crowded ships testified to the corrupt practices of pursers who made money by selling privileges to those who were willing to bribe, while persons who had paid for berths found themselves shut out unless they "took care" of those in brief authority.

As seen from the incoming ships the tented city on the

edge of the tundra looked like a snow-drift. Tents stretched from Fort Davis to Penny river, a distance of 22 miles.

Fully 2500 landed at Nome early in the summer of 1900 and camped on the edge of the tundra, where a white city five miles long faced the shore, then littered with freight and machinery, including some of the weirdest devices ever invented. Naturally "the golden sands of Nome" had served as a fascinating cry from irresponsible brokers to the gullible portion of the public. Companies were formed without limit and stock was

A RELIC OF THE BOOM ON THE BEACH AT NOME.

sold without stint, the flamboyant promoter spurning the handicap of truth or limited knowledge of such facts as might hinder him in framing his alluring prospectus. Of course, if individuals without machinery could earn $10 and $20 per day by mere digging of the sea-beach, it was obvious that with machinery and expert knowledge, the winnings would be tremendous. Thus they salted the tail of the bird of their imagination until they thought they held it fluttering within their greedy hands. Every kind of gold-saving device was brought to Nome, from patent cradles to cumbrous dredges. It was planned to dredge under the waves, the promoters believing that the gold was swept shoreward from the sea, and therefore the greater the distance from the shore the richer the sea-bottom. For this fallacy there was no excuse, because Brooks and Schrader of the Geological Survey had already published a preliminary report, giving a simple explanation of the character of the deposits. This is a fitting place to refer to the great service done by the geologists of the Survey, not to Nome only, but to Alaska in general, through the publication—without delay—of maps and descriptions of local geology such as the intelligent miner could use as a scientific guide in his daily operations. At Nome, in particular, the work of the Survey exemplifies the invaluable aid given by a scientific bureau to a young industry. The description of the beach deposits and the explanation of their mode of origin, published in the first reconnaissance reports, were in effect a prophecy that similar concentrations would be found on the coastal plain, and I do not doubt that the discovery of the Second and Third beaches was hastened by the elementary but fundamental geologic principles enunciated by Brooks and Schrader in 1900.

A few profited by the information published by the Survey, but to the mob it mattered little, and to the faker it was a hindrance. Therefore vast sums of money were squandered in the delirious trimmings of machinery, relics of which can still be seen cast away upon the shore. One fearsome machine is shown in the accompanying photograph; it had huge iron wheels with flat broad tires and carried a suction pump by which the sand was elevated to a washing apparatus. This contraption was supposed to stand in the breakers and reach the gold on the bottom of Bering Sea. Several dredges of nightmare design lie half buried in the sand. If there be few survivors of the array of machines that crowded the narrow beach in 1900, it is because a big storm on August 9 smashed them and swept them high on the tundra. A few linger on the winding estuary of the Snake river, and there I saw them standing as a warning, let us hope, to the inexperienced.

In June 1900 the *Oregon* brought four smallpox cases to Nome; when this fact became known there was a scare that

AN ESKIMO BELLE.

took the heart out of the mob of disappointed gold-seekers. Many had come expecting to find the beach glittering with nuggets. Some of these simple ones lost their reason. A man was seen to walk along the shore and reach down for a grab of sand, which he then spread on his hand; finding no gold, he blew out his brains. An old man and his wife pitched their tent on the beach and two days later were found dead; their disappointment had been too much for them, so the husband had first shot the wife and then himself. It was a tawdry settlement full of the flotsam and jetsam of the human tide. At one time as many as 2000 worked on the beach and the gold easily won was quickly spent by most of them in the saloons and dives that were so numerous. In the summer of 1900 there were 30,000 people at Nome, but 16,000 left in 13 weeks. They left hurriedly, selling their effects as best they could, so that one or two enterprising men bought supplies for two years ahead at prices that represented 10 cents on the dollar. The smallpox scare prompted many to decamp, the impending winter frightened others, inability to get work and the high cost of living made life impossible for even those who were willing to exert themselves, and these were shipped 'outside' either by the Government or by private charity. The output of gold in 1900, including the creeks, is estimated to have been $4,750,000. Since then this mining district has been the scene of anarchy disgraceful to the Government of the United States, but that is so complicated a story that I leave it for the moment, while we watch the life of Nome as it is seen today.

Nome, more than any other region, gave an impression of strangeness such as may well have befallen the adventurers of Queen Elizabeth's time when they first invaded the Spanish main. Here were uncouth simple natives, wholly unlike the American aborigines; here wood was so scarce that these natives treasured a piece of willow as if it were precious metal; here were no forests, only a dreary waste of tundra; here the air was so still that voices could be heard afar; here day and night were scarcely distinguishable, work proceeding at midnight as at noon in the ethereal illumination that most of the gold-seekers had never seen before on land or sea.

The Front street of Nome has more character than the thoroughfare of any other American mining camp. This it owes to the Eskimo. As Cairo is the meeting place of the Eastern and Western civilizations, so Nome is the spot where the people of the Arctic mingle with the invaders from the Temperate zone. This gives a strange diversity of costume. The gay cloak or *parka* is worn by both sexes; it is made of colored drilling,

ESKIMO GIRLS.

imported, of course; this is a new fashion, for the old Eskimo wore only the skins of animals, chiefly squirrel and reindeer hides. Reindeer-skin is supposed to be the warmest fur; the hair is close; it belongs to the animal living in the bleakest region of the globe. The undergarments—shirt and drawers—are made of the skins of reindeer fawns, the hair being worn against the body. The women wear the same clothing, but in the form of a 'combination'. In Siberia the natives wear a thicker garment of full-grown reindeer-skin. This undergarment, with the fur against the body, is covered with another having the fur outside; and this completes the costume, except that in traveling the Eskimo wears an overgarment that serves to protect him from the wind: a *parka* made of drilling, and provided with a wolverine hood. Wolverine is said to be the only fur on which the frost will not gather, as the long hair prevents it from matting. Other furs when dampened by the breath will freeze, forming icicles against the face. The Eskimo's socks are made of the hide taken from the legs of the reindeer; over this he wears the *mukluk*, a high leather boot, the body of which is made of hair-seal, the hair being sometimes removed. The top of the *mukluk* is made of reindeer hide taken from the animal's legs, and the sole is lined either with walrus or with seal-skin. In summer they wear a high waterproof covering made of the skin of the hair-seal, which is kept soft by the application of seal-oil, the smell of which emanates from everything belonging to the Eskimo. Rancid seal-oil has an odor that travels far and is unwearied. The reindeer-skins are soft and beautifully tanned, the inner bark of willow being used for this purpose. Deer-skin caps, with the hair outside and with no lining, are worn in winter. The women's heads are protected by their own hair, black, thick, and long. It is washed in an unmentionable fluid, but soap is now being used also. On first making the acquaintance of soap, they ate it, like the man who put his feet in the gruel, and drank the mustard and water. Mistakes will happen. The children's little *parkas* are fringed with a ruching of land-otter, and their oily brown faces, with a touch of deep red, are lit by intensely black eyes and a cheerful smile. They look happy, even though in clothing and manner of life they differ so much from the pink and white youngsters of the Southern folk.

On the sand-spit at the mouth of the Snake river an Eskimo settlement has been made. Most of the shelters under which they live consist of *oomiaks*, or family boats, turned on edge. A few tents give variety to the scene. Tom-cod in festoons are hung on poles, dogs are asleep in the sand, dirty native women

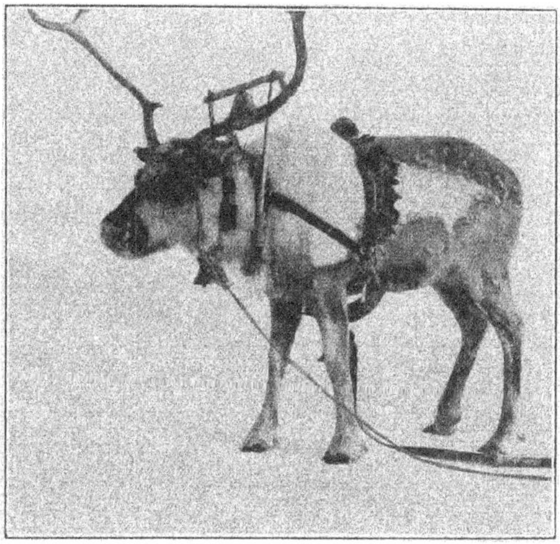

REINDEER.

AN ESKIMO IN HIS KAYAK.

crouch by the camp-fire, a young fellow is finishing a *kayak*, the tin débris of the boom days litters the shore, a smell of seal-oil is borne on the breeze—it is not an inspiring picture. These Eskimo live on Cape Prince of Wales and come to Nome in summer, to fish and to sell curios, such as carved ivory, mainly walrus tusks.

The Eskimo, or Innuit as they call themselves, live in underground hovels called *igloos*. To them also "there is no place like hime" and I can vouch for it that there *is* no place like the Eskimo's home. It is not possible to describe the habits of these people or their queer customs, for some of them would disgust the polite reader. They have been related in other books of travel. "One man's meat is another man's poison" and one people's ways would poison the minds of another. At night when at home the men strip nude and the women wear a breech-clout; as soon as they go indoors they remove their reindeer clothing, which, of course, is infested with insects that do not annoy them, mainly because their smooth hard skin is continually smeared with seal-oil. The clothes are hung on a pole above the lamp, fed with seal-oil, standing in the centre of the *igloo*, and thus they are dried. In frosty weather the native takes his skin clothes and beats them with two deer-horns provided with a crook, like a golf-stick. This they do the first thing in the morning so as to disengage the vermin.

On the Siberian side the natives rarely die a natural death. When old or diseased they are killed by hanging or by stabbing, often at their own request. Suicides are frequent, especially among the women. These have been known to go out into a winter storm nude and court death by freezing. Their lives are those of animals, and though they possess some sort of intelligence making them superior to the beasts that they hunt, it is a self-consciousness that only adds to their misery. I would rather be a seal or a polar bear than an Eskimo.

Come and dine with me at the Royal Cafe; it is not the Café Royal, and yet if previously you have walked over the tundra or ridden on horseback along the firm sands of Bering Sea, you will pronounce it a good restaurant, however you may pronounce the name it bears. The place is crowded but clean; the

ESKIMO CHILDREN.

POLAR BEAR AND HUNTER.

well intended efforts of a piano and a violin give a touch of gaiety, and the crowd that passes along the main street can be watched with interest while the reindeer stew or the roast ptarmigan is being prepared. The Eskimo give color to the scene; the women in their pink and yellow parkas and wolverine hoods look like ladies on their way to a party; the men in fur ruffles and light drill parkas wear visored caps or else go bareheaded with masses of long black hair trimmed with a Dutch cut. Two Eskimo carry the skin of a polar bear on a long pole. Others have carved whalebone for sale. Dogs are numerous. The bright tints of the native costume produce a chromatic liveliness unusual in a mining camp. The huskies and the malamutes accompany the Eskimo and suggest Arctic life. An occasional Saxon of fresh complexion looks very pink amid these black and oily denizens of the North. Stalwart miners in high laced boots and stiff broad-brimmed hats recall Colorado and Nevada. Women dressed conventionally indicate that Nome has homes as well as mines. The superintendent of a mine rides past on a handsome black horse that clatters over the boarded street and scatters the Eskimo children with the dogs. And all this time the musicians in the background have been doing their best, as well as the cook. Silver salmon, reindeer steak, and ptarmigan, followed by blueberry pie, represent adequate nourishment of a kind suited to the picturesque environment. A demi-tasse and a Havana cigar emphasize the fact that Nome is no jumping-off place, but on the highway of civilization from New York to Paris, viâ Bering strait.

CHAPTER XXX.

THE DOG RACE.

Eskimo and dogs make Nome lively, especially at night. A fire on Sesnon's wharf incited the dogs to a special outbreak of howling on one otherwise quiet night. The noise was something between the sad plaint of fiends in hell—so I am told—and the caterwauling of felines on the garden wall—this I know. The outcry was especially weird when it became faint, as if in hopeless agony. A few barks from 'outside' dogs could be heard clearly, but the uproar came predominantly from the native canines.

'Huskies' and 'malamutes' roam on every street and alley of Nome. *Huski* is a native word for Eskimo and *malamute* is the name of a tribe at the mouth of the Yukon. These dogs are stocky and sturdy, extremely muscular and long-haired, so that they are well fitted to pull sleds over the snow. When on the trail in winter the dogs are fed with dried salmon, one fish weighing two pounds being given to each dog. After the fish, the dogs are fed with three-quarters of a pound up to one pound of fat bacon, of the cheapest variety. Some men cook a mixture of salmon, rice or cornmeal, and bacon drippings. When cooled, this mess is arranged in little mounds on the snow at regular distances, so that the dogs may not poach on one another's allotment. They will eat anything fat, such as lard or tallow, the appetite for such food being developed both by man and beast through adaptation to a cold environment. If the bacon is not rendered out before being mixed with the cooked food, a clever dog will go from one pile to another and use his paw so as to pick out the scraps of bacon. This will lead to fighting, of course; for the dogs recognize no friend except the man who feeds them, and only while he feeds them. When fish is fed dry to the dogs, it is first thawed by being broiled slightly over a fire, so as to bring the oil to the surface. When on the trail the dogs are fed once a day, at the end of the run; if fed in the morning, they become torpid.

Travelers must take care not to be frost-bitten in winter when 'mushing' with a dog-team. In extremely cold weather, especially when windy, a patch of rabbit skin, with hair attached, is put on the tip of the nose and on each cheek-bone; this will delay freezing. The fur is moistened and becomes attached on the hairy side, the idea being that it protects the spots on the face where freezing would otherwise start. The natives wear wooden spectacles or a wooden eye-mask, with a slit for each eye, to guard against snow-blindness. If blinded by the glare of snow, a bit of raw meat or even a raw potato serves as a poultice to lessen the inflammation. That is a remedy used in the mountains of Colorado. When traveling in the North during winter, care must also be taken not to become so hot as to perspire, for if a man becomes warm and then stops to rest, the moisture freezes and he is in danger of a sudden chill, leading to pneumonia. The aim is to keep cool without becoming cold. When on the march the outer garments are doffed as soon as the traveler becomes warm, the experienced men usually wearing only a drill parka to protect them from the wind, leaving the fur parka for the time when they are weary or when their vitality has run down. Great care is taken of the feet, to keep them warmly clad. A man with frost-bitten toes in the wilderness of snow is in a bad plight. Ordinary socks, 'German socks,' straw in the sole, and then 'mukluks' over everything, constitute the best foot-gear.

The malamute dogs are miserable in summer, but full of life in winter. On the wildest night that blows a malamute will lie where the wind will strike him fair. They are extraordinarily hardy. By service with the sleds they develop into lively bundles of muscle; they enjoy their work in harness and run like spirited horses, so that the driver is compelled to apply his brake. The usual price of a good malamute is $50, although some of the dogs in the celebrated race brought $400 to $500 apiece.

The first dog-team race on the Seward Peninsula started on April 1, 1908, the course being from Nome to Candle and return, a distance of 440 miles. Ten competitors started. No limit was placed to the number of dogs in a team, but each driver had to return with the same dogs as he had at the start, whether alive in the harness or dead in the sled. This rule was meant to prevent cruelty or excessive strain on the endurance of the dogs. For instance, one dog in Bob Adams' team froze in his traces during the storm that swept over the country a

A DOG TEAM ON THE MARCH.

few hours after the start from Nome and his dead body was carried on the sled for the remainder of the race. The teams consisted either of seven or nine dogs, the odd one being the leader, who was an 'outside' dog, that is, a setter or St. Bernard of mixed breed. Intelligence and grit, rather than strength, are required in the leader, and for this reason the pairs of huskies and malamutes of the Eskimo follow the dog brought by the white man from another country. The harness includes a collar that is pulled over the head and padded, like a horse-collar. The dogs were fed with fish and bacon in the usual way, together with rice and condensed milk. Only one meal a day was permitted, preferably at night, after the day's work was over. The dogs in the winning team ranged in weight from 80 to 97 pounds, the average weight being 86 pounds.

The sled used in this race was made of light oak or hickory slats, shod with brass and overlapped with brass bands. The maximum weight of the sleds was 45 pounds, Fink's weighing only 20 pounds. Albert Fink is the Nome lawyer who owned the winning team. The race took 5 days, less 40 minutes, the first three teams arriving within 30 minutes of each other as measured by the actual running time, for they did not start simultaneously, but at intervals, so as not to interfere. The ten teams started two hours apart, the order being decided by casting lots, the winner choosing to start last, thinking it best to let the other teams break the trail over which he would follow. This helped Fink's team. For it so happened that the teams which started first felt the brunt of a storm that overtook them soon after the start, when only 40 miles from Nome. Six of the teams were held for 20 hours at Brown's road-house by reason of this blizzard. Thus the three teams that started last had a big advantage in time. And they tried to get every other advantage, coaxing the extra dogs running loose behind the sled in front of them, thereby hindering an opponent from reaching the next resting place, for each driver had to be sure of all his dogs, the absence of any one of them disqualifying the team.

The race was for a prize of $2500, and by private arrangement in one case, at least, the driver was to receive half the purse. But this did not measure the amount of money at stake, for betting was brisk. Fink made bets of $300 at 4 to 1 against the field in behalf of his driver and each of his three commissary men. The race was reported by the telephone service established by A. E. Boyd along the road between Nome and Candle; thus the arrival and departure of each team at any one of the thirteen telephone stations was reported, so that the entire race was watched from start to finish, amid the greatest local excitement. The bulletin board at Nome was never without a crowd, the interest being prolonged for five days. The winner made the course in 4 days 23 hours 15

THE TEAM THAT WON THE RACE AT NOME IN APRIL 1908.

Photograph by Dobbs, Nome.

Published by permission.

minutes, and the second team made it in 5 days 7 minutes 52 seconds, so that the difference in time over the 440 miles was only 52 minutes. These teams were delayed 20 and 22 hours, respectively, by the storm. The average time was 8 miles an hour. Of the three winning teams not a dog was injured by the race; in a couple of days afterward they were in harness again and ready for work. At one time it looked as if Fink would lose, so he hedged miserably. This was neither good sportsmanship nor good judgment, for he had a splendid driver—Bert Barber. On arrival he had seven dogs in harness and two running loose behind. The latter had lain in the sled for many miles—just tuckered out. From Candle to the Timber road-house this team traveled 140 miles without a stop and the driver states that he never got into the sled except to ride his brake when going down steep hills. The second team, driven by Jake Berger, made the 36 miles from Solomon to Nome in 4 hours, that is, at an average speed of 9 miles per hour.

The winner arrived after midnight. The whole town was agog and Front street was closely lined with the crowd for a mile long. The air was fifteen degrees below zero, but that was not noticed. The winning driver arrived hatless, without a coat, with his thin alpaca shirt thrown open, exposing his neck and shoulders to the freezing midnight air, arms bared above the elbows, braces hanging loose, his face red and perspiration rolling down, the steam of it being visible in the frosty atmosphere for three or four feet above his head—he was a sight! Completely exhausted, he seemed dazed when the long rush was ended. On his feet he wore mukluks, the native boots made of seal-skin and soled with walrus hide. His hands were bare, the gloves and fur-coat lying in the sled. It was a famous race, and the tale of it will long be told by many a camp-fire in the North.

CHAPTER XXXI.

THE THREE SWEDES.

The story of the gold discoveries that made Nome famous constitutes one of the great tales of mining. No flimsy romance is needed to decorate the facts and no mythological frills are required to enhance the results. The first discovery of gold made by white men was in 1888 when King and Green, who found the Omalik silver mine on Fish river, panned fine gold on the bars of Fish river. But they did nothing further to test the value of this placer. In 1894 Joe Hansen, who later made a fortune at Dawson, went with two natives up the Fish river, and then up the Niukluk, above Casadepaga, and finding more gold, came back for lumber to make sluice-boxes. Returning to the coast, he got word from a partner telling him of discoveries on the Klondike; thereupon he left for the upper Yukon. This Norwegian was named Johannson, a name easily corrupted by his American friends to Joe Hansen. He deserves to be properly recorded in the archives of the North, for not only did he make one of the first discoveries of gold, but he also taught the Eskimo on the Niukluk how to pan. That was a lesson fruitful in results; he deserves to be styled Professor.

Another important figure was John A. Dexter, a trader on Golofnin bay, who encouraged the natives to pan the gold of the creeks when on their fishing and hunting trips. Finally, in August 1897, an Eskimo, named Tom Guarick, found gold on Ophir creek and reported the fact to Dexter, showing him half an ounce of gold, which he had panned. Only a month later a party of prospectors came to Golofnin bay from St. Michael. They had come from California under a grub-stake agreement with capitalists at San Francisco. Being shown the gold at

Dexter's store, they secured Tom as a guide and were taken by him to the place on Ophir creek where this Eskimo had made a discovery.

This party of prospectors consisted of Daniel B. Libby, H. L. Blake, A. P. Mordaunt, and Louis S. Melsing, under the leadership of Libby, who knew something about the country, having been there with the Western Union Telegraph expedition in 1866 and 1867. When a young man of 25, he had charge of the Port Clarence station and in the fall of 1866 he had crossed the Seward Peninsula with Otto von Bendeleben. At that time they had detected the presence of gold in the gravel of the Niukluk river; when, therefore the Klondike excitement broke out he organized a party to test the value of the discovery made 30 years earlier. Landing at Golofnin bay, he was just in time to be told of the Eskimo's find and promptly utilized the information, in the manner narrated. Besides Ophir creek, they found gold in a neighboring valley called Melsing creek, christened after one of their own party. Other prospectors came on the ground, for the whole of Alaska was getting the benefit of the interest excited by the Klondike rush, and men continued to arrive from St. Michael. Thereupon, in accordance with the established custom of American mining regions, a district was organized and a Recorder duly elected on April 25, 1898. The district was named Eldorado, the organizers being the four prospectors already mentioned, as well as A. N. Kittleson, who had been in charge of the reindeer station at Port Clarence, N. O. Hultberg, a missionary of Golofnin bay, P. H. Anderson, a missionary-teacher from the same station, and John A. Dexter himself. As we shall see, missionaries and reindeer furnished local color to the romance of the Seward Peninsula.

The first exchange of gold dust for provisions was made by Nate Vestal, an old Montana miner working at the mouth of Sweetcake creek, and H. T. Harding, at Council, in August 1898. Council was the settlement established on the Niukluk at this time. About $75,000 worth of gold was won in the first season, but the news of the diggings on Ophir and Melsing creeks did not go far, so there was no stampede. At that time

CAMPBELL AND SAMUELSON ARRIVING AT NOME FROM VALDEZ ON APRIL 3, 1908.

wild stories of gold discovery had become so frequent that they had ceased to excite.

In June 1898, a party, consisting of J. L. Haggalin, John Brynteson, Christopher Kimber, H. L. Blake, and N. O. Hultberg left Council on an expedition, prompted by a report that coarse gold had been found on the Sinrock, or Sinuk, river by a reindeer-herder. Leaving Golofnin bay in a small boat on July 2, 1898, they were driven by a storm to take shelter in the estuary of the Snake river, close to the present site of Nome. This party of prospectors went up the Snake as far as Anvil creek and found some gold, but not enough to justify locations. They prospected on what was afterward No. 5 Below, but their work was of a desultory nature on account of a rain-storm then prevailing. This was on July 26. Returning to the coast, they went to the Sinuk, where nothing was found. So they retraced their steps to Council.

Brynteson and Hultberg were probably the first to find gold in the Nome district, for they found some gold at the time when the first party went up the Snake, and they related the fact on their return to Golofnin bay, where Hultberg was stationed as head of the Swedish mission. Another result of the unsuccessful expedition was that Brynteson determined to go to Anvil creek later in the same year, accompanied by Lindeberg and Lindblom. Here we turn to a page vivid with human achievement. Lindeberg, Lindblom, and Brynteson are known as "the three lucky Swedes" and the story of their discovery of gold, with the litigation that followed, is the iliad of Nome.

John Brynteson, a native of Sweden, was an experienced coal and iron miner, who had worked for seven years in the iron mines of Michigan. Determining to go to Alaska and search for coal, he reached St. Michael. Shortly afterward he directed work at the coal mine on Norton Sound operated by the Swedish mission at Unalakleet, then in charge of Haggalin.

Erik O. Lindblom, another Swede, for several years a tailor in San Francisco, hearing of fabulous gold discoveries on Kotzebue sound, joined in that stampede and came north on the bark *Alaska*. Arriving at Port Clarence, and hearing of the

ON THE BEACH OF NOME IN WINTER.

Ophir creek excitement, he left the ship and found his way by the coast to Golofnin bay, and thence to Council.

Jafet Lindeberg, a Norwegian, came to Alaska as one of the men in charge of the reindeer herd brought by Dr. Sheldon Jackson at the instance of the United States Government in 1898. He was to have gone to Plover Bay, in northeast Siberia to relieve Captain Kelly, who was trading at that place for reindeer in behalf of the U. S. Government. Arriving at St. Michael, news came to Jackson that Kelly had been driven from Plover Bay by hostile natives, whereupon, it being deemed unwise for Lindeberg to go to Plover Bay, he was released from the service of the Government. Then he also went to the new diggings on the Niukluk, going first to Council.

Thus these three men chanced to meet at Council City in August 1898. Brynteson was then 40 years of age, Lindblom was 30, while Lindeberg was a young fellow of 21. While prospecting in the vicinity, they also learned the mining laws; for by that time the district was organized and had established regulations, one of which was the right of location by an agent or attorney in fact. Going to the Ophir diggings, they found the creek pretty well covered with locations but not much work was in progress because the gravel was considered rather poor. They panned enough to get an idea of the yield requisite to make operation profitable. Harry L. Blake offered to let them work and take $20 per day, giving him 25% of any surplus. This was on No. 4 Above. But they had formed a "prospecting companionship" (as Lindeberg phrases it) and decided to search for gold over a wider territory, where there were fewer people. The Council district was now over-run by 'stampeders' and the country was staked "to the mountain tops." Returning to Golofnin bay, the three procured a large open boat, stocked her with provisions, and set sail on September 11, 1898, on a quest that proved eventful. Proceeding up the coast they stopped at the mouths of the various rivers in order to prospect, but finding nothing noteworthy they reached the site of Nome, where the Snake river flows into Bering Sea. Noting its serpentine channel, they named it appropriately. Ascending this stream in their boat, they finally camped at the mouth of

Glacier creek, where Brynteson had found gold on the occasion of his previous journey to this region. Then the three tested the various creeks, including Anvil, Snow, Rock, Dry, Dexter, and Glacier, besides several (such as Sunset and Buster) creeks

A TEAM OF HUSKIES.

on which they made no locations. It is one of the remarkable facts in mining history that the hasty exploratory work of these three men, comparative novices as they were in prospecting for gold, should have resulted in the selection of what proved later to be the richest portions of the several creeks.

Lindeberg states that they found gold in payable quantity on Anvil, Snow, Glacier, Rock, and Dry creeks. He says: "We proceeded to locate claims, first on Anvil creek, because we found better prospects on that creek than on the others; therefore we located the 'discovery' claim there in the names of us three jointly. In addition to this, each one of us staked a separate claim in his own name on the creek. This was the custom in Alaska, as it was conceded that the discoverer was entitled to a discovery claim and one other."

And they located with rare judgment. The three original claims staked by them on Anvil proved to be the best. On Dexter they staked No. 8, 5, and 3; of these, No. 8 was the best of the three, although not the richest on that creek. No. 5 and 3 also proved good. On Snow gulch they got No. 1, 2, and 3; these three claims proved the only rich ones on that creek. On Rock they located No. 2, 3, and 5—undoubtedly the best on that creek. On Dry they got No. 3, 4, and 5, all above discovery. The Discovery claim was located by G. W. Price. For the present it suffices to compliment the three on their skill as prospectors and to record the fact that all the claims they located were subsequently consolidated under the name of the Pioneer Mining Company of Seattle.

The missionaries were not without enterprise. The three located No. 9 Anvil in the name of an Eskimo. To ensure title the claim was re-located in the name of Gabe Price's brother and by him deeded to P. H. Anderson, then in charge of the Swedish mission on Golofnin bay. He took out about $800,000 gross or about $500,000 net, but he claimed the gold for himself. Judgment was obtained by the Mission against him for $250,000, it being held that he had the property in trust. Another incident of a more creditable kind was the locating of No. 15 Ophir by Carlson, head of the Unalakleet mission; he sold the claim for $5000 in behalf of the mission. Later this claim yielded over $1,500,000. It will be noted that by reason of Swedish missionary work among the Eskimo and the introduction of reindeers by the Government, the Scandinavians figure largely in the story of gold discovery on the Seward Peninsula. Many of them were not naturalized citizens of the United

UNLOADING FREIGHT FROM THE CORWIN, OFF NOME.

States, and this fact was used as an excuse for an attempt to despoil them of their mining claims, leading to an era of gross political and judicial jobbery.

On the return of the prospectors to Council, in October, the news quickly spread and caused a rush. A party was organized by the three Scandinavians, together with A. N. Kittleson, G. W. Price, and P. H. Anderson. Upon their arrival at the mouth of the Snake river, on October 18, a meeting was held, the Cape Nome district was formed, and A. N. Kittleson was elected recorder. Although the summer season was now ended and it was therefore too late for surface mining, there was time to stake claims. This was done without limit. That iniquitous fiction known as 'power of attorney' was used so recklessly that 7000 acres of rich placer ground was staked by not more than 40 men. Only about 30 claims were located by the original discoverers, for themselves and for their friends, and by the second party, consisting of Kittleson, Price, and Anderson; but the reckless use of powers of attorney by the party from Council City, or the Eldorado district, soon plastered the region with locations; these last not only re-located every claim that had been pegged up to that time, but they located on new creeks without doing any work to ascertain the presence of gold. When Blake, Mordaunt, Libby, Melsing, and the others who had done the first gold mining on the Peninsula learned that the 'three Swedes' had located the best claims, they were chagrined, for the Scandinavians were inexperienced in gold mining as compared to the prospectors whom Dexter's Eskimo retainer had led to Ohpir creek. Thereupon they jumped the claims of Lindeberg and his partners on Anvil creek and thereby set an example that was promptly followed by the crowd of newcomers attracted by the excitement. In consequence, every original claim was covered two and three deep by re-locations, preparing the way for endless trouble and litigation. This ended in anarchy.

CHAPTER XXXII.

THE GOLDEN BEACHES OF NOME.

To dig gold from a sea-beach seems so simple a form of mining as to be highly improbable, and yet that was the motive to a strange drama enacted on the shore of Bering Sea in 1900.

Natives had detected gold on the beach long before the white men came. They had reported the fact to Kogan, the captain of a whaling ship, who traded with the Eskimo living on Cape Prince of Wales; but he paid no attention to the story. In August 1898, a prospector named Tom Mulligan found gold on the shore at a place half a mile east of the mouth of the Sinuk river, 30 miles west of Nome. He found enough to warrant the belief that he could make wages, that is, $8 to $10 per day. But the Anvil creek discoveries diverted him. His discovery was made known to the thirty men who were camping on the Sinuk that winter, and they planned to work the place during the following summer, but the wonderful richness of the fringe of sand in front of Nome led to the abandonment of their plans.

On June 28, 1899, a soldier found gold on the beach at Nome. He belonged to the small representation of the United States army brought thither by the threat of impending disorder. This soldier used to pan enough gold to pay for an extra meal, but the first "big money" was taken out of the sand by William Fee, otherwise known as Missouri Bill, and his partner, William Cummins, both old Yukoners. This was on August 1. The astonishing fact soon became known. All the idle men, unable to find work owing to the jumping of claims on the creeks and the impending litigation, hastened to construct rockers and wash the golden deposit on the shore. By the tenth day of August fully 1500 men were at work on the beach; from being 'broke,' they were winning from 2 to 10 ounces of gold per day. "It livened the town considerable." It is said that the highest yield from a day's work with a rocker was 129 ounces of gold, which was a clean-up made by Missouri Bill and his partner. Within three days silver coins disappeared from circulation, for the dollars were used in the process of saving the gold by means of amalgamation. Gold dust became the sole medium of exchange. Mercury went to $5 per pound. All the sheathing on the boats and the big coffee urns in the restaurants were utilized for the copper needed in the rockers and long-toms. Lumber being worth $400 per thousand, the diggers constructed rockers out of any sort of material, such as the boxes in which the condensed milk was packed. In less than two months 2000 men extracted over $1,000,000 from the beach.

The method of mining was simple indeed. First the prospector tested the sand by washing it in a pan. If the result indicated that he had found a rich spot, he started to dig a hole, from which he obtained a supply of gold-bearing material. This he fed into a 'rocker' or 'long-tom,' two devices of early origin. The 'long-tom' essentially consists of an inclined surface over which the gravel or sand is washed by water fed by hand. From 6 to 10 feet of launder or sluice-box is set at an angle steep enough to permit the light particles to be washed away while allowing the gold to settle on the bottom. At the head, a hopper or box serves to hold from 50 to 150 pounds of material which is flushed, a little at a time, down the slope, by the action of water thrown out of a dipper or small bucket fixed to a handle. To arrest the gold, cross-bars or riffles are nailed to the bottom of the launder. In addition, mercury may be employed, or even amalgamated copper plate. When the latter is used the plate is covered with wire screen or perforated sheet-iron, the effect of which is to size the gravel, causing the larger pebbles to slide down the slope, while the fine stuff sinks through the apertures and comes in contact with the mercury and amalgam. The 'long-tom' was familiar to the early Californian miners, and in its simplest form dates back to the very beginning of the world-wide search for gold. In order to facilitate the process of concentration, a shaking motion was imparted, merely by placing the inclined sluice-box or launder upon rockers. This constituted the 'cradle,' which is shorter and more compact than the 'long-tom,' the quicker separation of the gold rendering unnecessary a long surface.

These methods are still in vogue. It so happened that during August 1908, while I was at Nome, there was a sudden increase of beach mining. At one time I saw more than 100 men at work. The beach is steep and forms a fringe only 60 yards

WORKERS ON NOME BEACH, 1908.

wide between the tundra and the tide. The scene of greatest activity in 1908 was in front of the town itself, under the wharves and in the rear of the houses. As the beach is a Government reservation, no location is possible, each man holding a small patch of ground only as long as he works upon it. Where the beach had been found most productive, the long-toms were thick and the workers numerous, but without any suggestion of disorder. Each man knew his rights and forbore to trespass. The apparatus varied according to the means of the operator. Most of the long-toms had a false bottom of galvanized iron or else tin-plate torn from oil-cans. This is punctured with holes so as to act as a screen separating the fine stuff from the coarse; the latter runs down the slope into the sea, on the edge of which the apparatus is erected. The fine sand, including the particles of gold, drops through the false bottom onto amalgamating plates. These are of copper, usually silver-plated. At the end of the copper plate there is, ordinarily, a bit of carpet, matting, or wire netting to serve as a check on any gold or amalgam escaping from above. At the upper end of the box or launder, the sides are raised or a hopper is constructed; into this the sand is discharged from a bucket, emptied from a wheelbarrow, or shoveled direct from the ground that is being exploited. While one operator attends to this part of the work, his partner is furnishing the water to wash the sand down the incline, standing with rubber boots in the tide and swinging a ladle consisting of a bucket fixed to a long wooden handle. Sometimes, for convenience, a temporary dam is made with bags of sand, forming a pool which is renewed by the waves that break over it at intervals. Some of the contrivances that I saw were pathetically crude; in one instance a small strip of old carpet and a few globules of mercury constituted the entire gold-saving system.

A more systematic arrangement commonly seen on the beach is a series of 6 to 10 sluice-boxes, set on a slope so that the reject falls into the sea and is removed by the tide. The boxes are lined with woven wire, having 2 to 4 holes per square inch, lying upon the cocoa matting that covers the bottom. Mercury is sprinkled on the matting by shaking it from a bottle through a cloth stopper. In order to expedite operations a small gasoline engine works a pump to supply the water for washing. The intake pipe of the pump rests on two wheels so that it can be withdrawn during stormy weather and the mouth of the pipe is protected with a wire net to shut out drift-wood.

The distribution of the gold is erratic, so that constant panning is necessary in order to ascertain whether it is rich enough to be profitably worked. Thus I saw a man, shovel in hand, scrape the top sand to one side and then dig into the red layer underneath. Shoveling about 20 pounds of this stuff into his

ON THE BEACH, NOME, 1908.

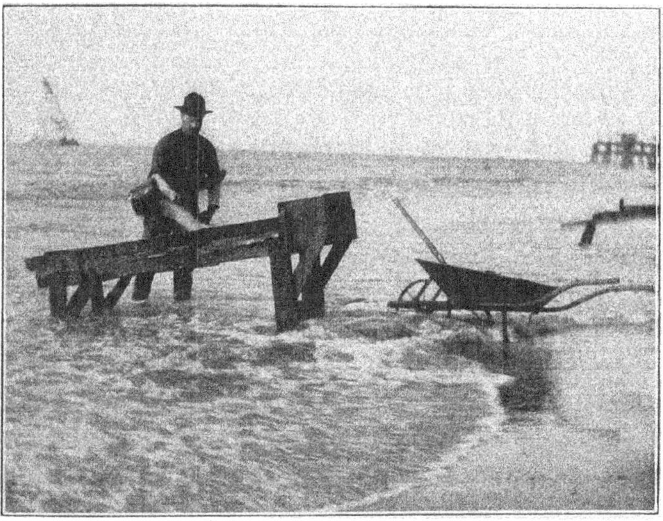

WASHING GOLD-BEARING SAND.

pan, he walked to the edge of the sea, dug a hole, which, becoming filled by the incoming wave, served as a basin in which he panned the sample. He obtained three "good colors," that is, three particles of gold worth about 10 cents. This was the first gold I had ever seen won from the sea-shore, although similar deposits are known in Oregon and California, in Tierra del Fuego, and also off the west coast of New Zealand.

The gold in the beach at Nome is the result of a natural process of concentration, in which the surf is the final agent. The gold-bearing material thus concentrated is derived from the rocks of the coast, the gold occurring in small quartz veins in soft schist, which is the prevailing formation. As the schist is weathered and eroded, the softer portions are swept into the streams and carried by them far out to sea, while the shattered quartz yields particles larger than silt but small enough for transport by running water. When borne to the mouths of the rivers this gold-bearing quartz gravel meets the tide and the surf, and by them it is washed to and fro, until the heavier particles are thrown in a narrow band at the upper edge of the beach where it is topped by the tundra. Among the heavier particles thus deposited is the gold, which, by disintegration of the quartz that contained it, has been released and now in the form of flakes of metal lies concentrated in a fringe along the 30 miles of Arctic coast.

In the hills four miles north of Nome are found quartz veins, carrying gold and traversing the soft schist similar to that which in a former period yielded the material for the beach placer. Between these hills and the beach, a coastal plain extends, flat and undulating, crossed by several meandering streams in the beds of which gold-bearing sand is found. This coastal plain is covered with the tundra or Arctic moss, mantling a deposit about 100 feet thick of gravel and sand, all of which is gold-bearing, although not all of it is rich enough to be mined. Only where concentrated by the running stream or by the sea is the deposit enriched. Under the deposit is the rock, either soft schist or limestone, similar to the formation observed on the hillsides to the north. Approaching the sea the surface of the plain slopes gently until it ends in an escarpment or abrupt slope only 10 to 15 feet high. At the foot of this declivity the beach slopes to the sea at an angle of 4 to 5 degrees, and for a width of 50 to 75 yards. During stormy weather the action of the waves extends for the full width of the beach, but ordinarily such action is restricted to the lower half of the shelving strand.

The beach is composed of sand and shingle, in which both the quartz and the schist are easily recognized. At the limit of the tide a reddish band is noticeable. This is the celebrated

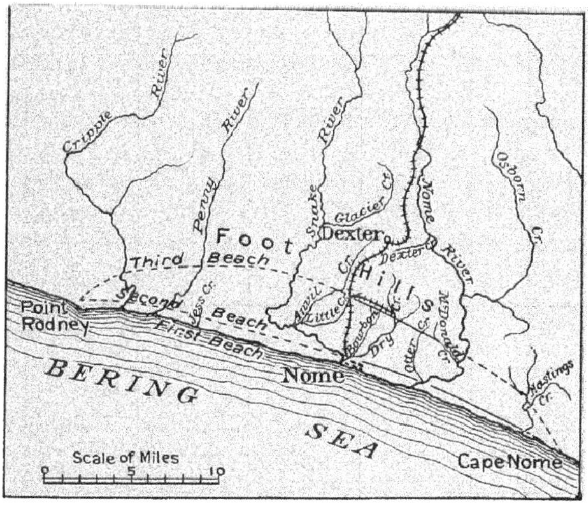

THE GOLDEN BEACHES OF NOME.

'ruby sand' of the miners. On examination it is found to owe its color to particles of garnet, with which a little black magnetite is also present, darkening the tint. At the foot of the tundra, that is, at the upper edge of the beach, a layer of clay can be detected, dipping under the ruby sand. This clay serves as a 'false bottom' under the gold-bearing garnetiferous sand, and upon it the valuable concentrate has been deposited.

By digging a hole into the beach, it can be ascertained that below the covering of barren gray shingle and sand is a layer of ruby sand lying on clay. The clay is from 3 to 7 feet deep; the gold-bearing red sand is from 6 inches to 2 feet thick; the fringe that represents the marine concentration is from a few inches to three or four feet wide. Sometimes two gold-bearing layers exist, separated by nearly barren sand. As the deposit lies on the bed of clay the dip is toward the sea. The flakes of gold are small, the largest weighing one pennyweight; these particles of metal are bright and amalgamate freely, although the grains of quartz associated with it are iron-stained. About $5,000,000 has been won from this beach deposit.

In 1899, the beach-workers got as much as $5 to $10 per pan; and even with the roughest contrivances, of the kind already described, some individuals in one summer season of only four months took $30,000 to $40,000 from the diggings on the shore. Today a man can still make $3 per day on the Nome beach. Two partners told me that they had made $60 in 3 days. Another operator and his partner got 3½ ounces of amalgam, yielding a little over an ounce of gold, on the day previous. Storms re-concentrate the sand repeatedly; the appliances required are cheap and easily constructed. It is a poor man's mine.

CHAPTER XXXIII.

ANARCHY AT NOME.

No account of Nome is complete without detailed reference to the anarchy that prevailed in 1900, 1901, and even later. Owing to the lax administration of law, due to the unorganized condition of the community and the great distance from the seat of national government, the titles to mining claims were disputed soon after the three Scandinavians and their friends had located the gravel on Anvil, Dexter, and other creeks that proved so productive. In the first place, the method of making locations was careless; it is true that the claims located in 1898 by the discoverers and the party who organized the Nome district were marked with six stakes, and although such stakes were small, being made from willow brush, they indicated an intention to conform with the legal requirements. But this good example was not generally followed; it soon became the custom, in marking a claim, to take a sprig of scrub-willow, blaze it, cut off the top, and split it. In the cleft was lodged the paper giving notice of location; then two end stakes were placed 1320 feet apart, up and down stream, fixing the centre line of the claim; from this line 330 feet on each side was assumed, so that an area of 20 acres was delimited. No stakes were placed at the corners. This perfunctory procedure became the recognized custom early in 1899 and was respected on account of the lack of such timber as was needed for complete legal marking of boundaries. Other than the scrub-willow there was no possible source of timber except the drift-wood on the beach four or five miles distant. During the summer of 1899 prospectors began to mark their corners with sticks of crooked willow and even drift-wood; but these, being stuck into frozen ground, were apt to fall down, and as they were rarely supported by a mound there was a tendency for boundaries to become questionable, especially when the pencil-marks on the location-notice became obliterated by the sap of the wood enclosing the paper.

In September of that year the big rush to Dawson, in Canada, caused many locators to depart for the Klondike diggings, and in their absence the claims were 'jumped' by newcomers, whose title in turn was disregarded by later claim-jumpers. Thus extreme confusion of title prevailed. Most of the people attracted by the excitement on the creeks near Nome were ignorant of mining and of mining regulations; yet some of them had sufficient education to undertake to interpret the law against the Scandinavians and others having no experience in such matters. A few really experienced miners, disgusted to see the best ground gobbled by the lucky Swedes and Norwegians, sailors and reindeer-herders, took it upon themselves to over-ride the law, as they themselves knew it, by jumping some of the best claims on Anvil creek. This gave a lead to the mob that was then arriving from the States. Among them were many lawless individuals and, worst of all, a number of unscrupulous lawyers. These, uniting, soon brought Nome to the verge of riot. A few Northwest Mounted Police or an honest Commissioner backed by his Government, as at Dawson, would have put an end to the trouble; but Nome was not Dawson; Nome was in Alaska, neither a State nor a Territory, but a 'district' governed from Washington, 5500 miles distant; the nearest United States commissioner was at St. Michael, more than 100 miles away; there was no means of enforcing the civil law. It is true a small detachment of soldiers had been sent from St. Michael in the spring of 1899, and upon the young lieutenant in command of this handful of soldiers devolved the duty of maintaining order among a crowd of angry men; but he could not do much, for he had no legal authority. Unfortunately, he exceeded what little authority he had by dispersing a proper meeting of miners and thereby permitted the idea to spread that he had taken sides. Finally, on July 13, the military issued an order stating that all disputes over claims were to be brought before the civil authorities, neither disputant being allowed to do any work pending a settlement.

NOME IN 1899.

Photograph by U. S. Geological Survey. By courtesy of A. H. Brooks.

This only made matters worse, for it meant that practically every claim would remain idle and thousands of men would be thrown out of work in a region where the cost of living was abnormally high. Moreover, it was unfair to many bona fide owners who had been victimized by irresponsible claim-jumpers. To meet the demands of the situation, the military order was modified.

But relief came suddenly in another way. In June gold was discovered on the beach at Nome and the mutterings of discontent were silenced in the excitement of profitable digging along the water-front. According to United States law a strip 60 feet wide of the shore is the property of the Government. Thus there was no title save that of Uncle Sam. A company, which had obtained control of a number of claims on the edge of the tundra so as to cover the beach, claimed ownership and proceeded to exact a royalty of 50 cents per day for the right to mine on the sea-shore. The miners objected, the military were invoked, and several hundred men were marched off the beach. But there was no magistrate to try them, no jail to hold them, and no funds to provide for them pending a trial, so the perplexed officer in command of the troops had to release them; whereupon they returned promptly to the work on the beach.

During this time the trouble on the creeks simmered, pending further developments. In the winter of 1899 the matter was brought to the attention of Congress and on June 6, 1900, an act was passed creating a new judicial district for the Seward Peninsula. The newly appointed judge and his court officials arrived at Nome early in the summer of 1900. Here we must leave the current of events in order to make a few necessary explanations.

The mining law of the United States requires the discovery of mineral on the ground located; in most cases the gold lay on bedrock at a depth of ten or twenty feet beneath the surface of the gravel; and yet some men staked a dozen or more claims in the course of two or three days. Moreover, none but citizens or those who have declared their intention to become citizens have the right to make locations; yet many of the richest claims were taken up by aliens, but in such cases the question of ownership could not be raised against the claimant by anyone except the United States government, whose land was thus pre-empted. A citizen has no right to jump a claim located by an alien until such time as the Government ejects the latter. Furthermore, many claims were taken up under the cloak of that legal fiction called 'power of attorney,' by virtue of which a man locates half a dozen claims in the name first of himself, and then of his father, brothers, and mother-in-law's relatives,

A MINE ON THE TUNDRA, NEAR NOME.

until a whole creek deposit has been blanketed. On top of all these complexities add unscrupulous lawyers eager to make matters worse, and it is easy to imagine what a devil's cauldron of litigation was brewed.

In the next scene the chief actor is a corrupt judge and his satellites, backed by equally corrupt United States senators, under the leadership of one of those predatory financiers who have been the curse of American industry. It was a gigantic conspiracy, the inner workings of which were finally dissected and exposed in the courts.

During the early part of 1900, while Congress was engaged

in drafting a code of laws for Alaska, the ownership of many of the richest placer claims in the Nome region was in question, and many lawsuits over titles were pending, owing to the fact that a majority of the best claims had been originally located by illiterate aliens; Nome was far away; here was a great opportunity for those to whom legislation is a financial game. Three local lawyers saw the chance; one of them went post-haste to Washington. There he enlisted the cupidity of several senators and one political boss, Alexander McKenzie, a Scotch-Irishman who had risen to wealth and power in North Dakota, becoming Republican leader in the days when Mark Hanna dominated McKinley. Hubbard, a lawyer from Nome, and Chipps, who had jumped one of Lindeberg's richest claims, joined with McKenzie in organizing the Alaska Gold Mining Company, capitalized at $15,000,000 under the laws of Alaska, 49 per cent of the stock being set aside to pay for the so-called title of the 'jumpers,' while 51 per cent was intended for distribution among McKenzie's political and financial 'friends,' by whom the nefarious scheme was to be backed in Congress and supported on Wall Street. The plan was to obtain title to the rich claims, which had been jumped, either by act of Congress or by decisions of the local court established at Nome and under the thumb of the gang. Even if the rightful owners fought against confiscation, the mines would be placed in a receiver's hands, the receiver to be appointed by the judge whom McKenzie had hired, and thus while the litigation was pending the gold would be garnered and sent to New York for exhibition with a view to selling the stock of the Alaska Gold Mining Company at such prices as would enrich all concerned. In order to accomplish this purpose the judge at Nome had to be perfectly pliant. He was. His name was Arthur H. Noyes. "He did the rest." If the scheme failed it was because a few resolute men fought like a moose at bay, fought so courageously and persistently that the scandal was finally exposed.

A bill providing civil government for Alaska was passed by Congress on June 6, 1900, and at the same time a code of laws was enacted, under which the district judges were given unusual powers. This bill as originally reported to the Senate provided that the laws of the United States relating to mining claims and mineral locations should be extended to Alaska. Numerous decisions of the courts had held that under the laws of the United States the citizenship of a locator of a mining claim could not be questioned by anyone save the Government; thereupon, while the bill was under discussion in the Senate, Hansborough, the senator from North Dakota, proposed an amendment to make it illegal for aliens to locate or to hold mining claims in the District of Alaska, thereby permitting

IN A DRIFT MINE.

litigants to raise the question of alien ownership. The amendment was also made retroactive, declaring null and void the title to claims located by an agent or attorney in fact. Finally, by declaring illegal all transfers of claims so located, the amendment destroyed the property rights acquired by purchase. This last little 'joker' was aimed at Charles D. Lane, who had acquired many claims, and was one of the leaders of the opposition. Lane was at the head of the Wild Goose Mining & Trading Co.; Lindeberg was president of the Pioneer

Mining Co.; these two companies had consolidated the richest claims on Anvil, Ophir, and other notable creeks. Several men hastened from Nome to Washington to explain what the effect of this enactment would be. Of course, it was unconstitutional, but that did not deter the conspirators nor help the rightful owners, for before it could be declared unconstitutional by the Supreme Court, the mines would be gutted. Additional proof of the conspiracy at Washington is afforded by the following fact: When the Senate committee on Territories had under consideration the law regulating appeals from the Alaska courts, Senator Carter insisted that a provision in the law allowing appeals from orders appointing receivers should be stricken out, and this was done, over the protest of two attorneys from Alaska who did not represent the Nome country and knew nothing of the conspiracy. However, later, a special statute allowing such appeals was, at the very end of the session, passed by Congress. Evidently the conspirators were caught napping. Fortunately, the dangerous character of the Alaska legislation was appreciated by several leaders in Congress, such as Spooner, Nelson, Jones, and Stewart in the Senate; these men obtained sufficient support and the amendment was killed. Thus the mining regulations of Colorado, California, and the other States became applicable to Alaska, and McKenzie's gang had to depend upon over-riding the laws by means of a corrupt judge. The scene was shifted to the court of the Second District at Nome.

On July 19, 1900, Noyes and the officials of the Court reached Nome. On the same steamer was McKenzie. Among his schemes was the Golden Sands Mining Company, which claimed title to the shore and tried to eject the beach-workers, as already described, under an interpretation of the law given by Noyes in open disregard of the statute. On July 23, before the Court was organized and before the filing of any paper of any kind with the clerk of the Court, Noyes appointed McKenzie receiver of a number of mines the titles to which were disputed, giving him instructions to take immediate possession, with permission to operate the mines, and hold the proceeds subject only to the Court's order. The defendants were ordered to yield possession at once and were enjoined from interfering with the management of the mines. In each case the receiver's bond was fixed at $5000, although the output of gold from any one of the mines was worth more than that in a single day. All this was done in impudent disregard of the law, without citing the defendants, who were taken completely by surprise. As the jumpers' titles were largely vested in McKenzie by this time, it is not necessary to comment on the Judge's action in appointing him, one of the litigants, as re-

NOME IN WINTER.

ceiver. The next day the lawyers for the Scandinavians and for Lane appeared before Noyes praying for a hearing. He refused. They then asked for an appeal; again he refused. "The Court claimed, in effect, that its jurisdiction in the matter was exclusive." All this took time. Meanwhile McKenzie had hired all the men available and was gutting the mines. On July 25 Noyes issued a further order empowering the receiver and his hirelings to seize everything on the mines, inclusive of personal property and gold extracted from other mines. Concerning this order the Circuit Court of Appeals at San Francisco stated, at a later date, that it was "so arbitrary and unwarranted in law as to baffle the mind in its effort to comprehend how it could have been issued from a court of justice." But as Alaska was neither a Territory nor a State, and there being no official higher than this Federal judge, who promulgated edicts like those of a South American dictator, the people of Nome had a taste of anarchy in America. It was the complete subjection of law and the domination of piracy.

Meanwhile the nearest Court of Appeals was 3000 miles away and the gold was near. McKenzie or one of his men was made receiver under twenty different injunctions, the owners of the claims being refused permission to be present at the clean-ups or to exercise any sort of protection against wholesale fraud. Nor did they stop at Swedes; a native of Ohio was dispossessed on the allegation of his being an alien. McKenzie became too bold, he boasted of his backing at Washington, he hinted at the support of the biggest men in public life, he suggested that the looting of the land was under government protection. It became a national scandal. Exploration and development of mines practically stopped. Miners were afraid to work profitable ground lest McKenzie and his gang should annex their claims. Local industry was paralyzed as by a Central American revolution. But the United States is not Guatemala; at length the higher Court at San Francisco granted the appeal that Noyes had denied. McKenzie was commanded to cease all action in the suits and turn back to the defendants their mines and other property. Noyes was directed to stay all proceedings in his court, the matter being taken out of his jurisdiction.

On September 14, 1900, the papers for the defence reached Nome; local feeling ran high; McKenzie was threatened with a violent death, but he stood his ground, refusing to recognize the writs or deliver the gold, and Noyes declined to act in accordance with the directions of the upper Court; instead, he called upon the troops to guard the bullion in the bank. Thus the Judge employed the United States Army to prevent the enforcement of the orders issued to him by the higher Court. It was the very burlesque of law, a travesty on civilization, the

WINTER DUMPS AT LITTLE CREEK, NEAR NOME.

complete conquest of graft. There was nothing to do but to send the news to San Francisco and wait for further action. This meant another month's delay. And all this time McKenzie and his gang were gutting the mines. But Noyes was careless and timid, he rendered no written opinions and took no pains to lend an air of regularity to his court procedure; even the records of his court, with the affidavits of the defence, were sufficient to convict him. Noyes became scared. It was about time. On October 15 two deputy marshals sent by the Court of Appeals landed at Nome with instructions to produce the *body* of the man who was charged with contempt. McKenzie accompanied the deputies to San Francisco, but before they left they had to break open the Alaska Banking & Safe Deposit Co.'s vault, in which he had placed his gold. Gold dust to the amount of $400,000 was removed from the vault to the Alaska Commercial Co.'s office and was subsequently returned to its owners. On the last boat to leave Nome, late in October, were the certified copies of the Court records; these had been secured by Samuel Knight, whose departure Noyes tried hard to prevent. The trial of the contempt cases took place at San Francisco. On February 11, 1901, McKenzie was sentenced to one year in prison, but he remained in jail for a few weeks only and was then pardoned on the score of ill health. He was then a member of the National Republican Committee and his pardon was obtained by the personal request of President McKinley. At that time McKinley happened to be in California; while at Monterey he telephoned to Judge Morrow, of the Court of Appeals, expressing a wish to see the latter. On the arrival of the Judge, the President stated that never had so much pressure been brought to bear in favor of a convicted person as in this McKenzie case, in fact, he doubted whether, if the positions were reversed, he himself could command as much political influence in his own behalf; and while, of course, he would not for a moment be a party to allowing wrong-doing to go unpunished, yet in this case, etc., etc.; whereupon McKenzie was promptly released. This is an excellent example of the submission of justice to politics; it illustrates the system under which the nomination of judges is the perquisite of the political boss. Although in Federal cases the President is vested with the pardoning power, McKinley preferred to act on the recommendation of the Judges because the charge against McKenzie was contempt of court. Morrow had served in Congress with McKinley; the Judge needed no explanations.

But conditions at Nome continued bad. Injunctions were issued freely; corruption was rampant. Noyes became more vacillating and more drunken. He reversed himself, he rendered no written opinions, he delayed decisions; then the long

A MALAMUTE TEAM.

winter came, there was no communication with the outer world, Nome was helpless. At one time preparations were made to hang Noyes and the court gang, and in the vigilance committee then organized were most of the leaders of the community, not ruffians but professional and business men of character. Just at this time the Judge was cited to appear before the Court of Appeals to answer for contempt. He decided to go. He was found guilty, and fined $1000 in lieu of imprisonment. Two of his assistants, named Frost and Wood, were imprisoned. Frost was the special examiner from the

Department of Justice sent to Nome by the Attorney General, but easily captured by McKenzie, who 'loaned' him money. Wood was U. S. Attorney and one of McKenzie's appointees. Political appointments at that time and by the gang then in control of the American administration were made in cynical disregard of decency. When someone protested against one of the appointments to the Court at Nome, Senator Carter replied: "He was a good man to get out of Montana, wasn't he?" An affirmative was inevitable. Even Griggs, the Attorney General of the United States, backed Noyes and wrote to him, in the summer of 1900, commending his acts in appointing a receiver under the circumstances related.

Judge Ross, who delivered the opinion in these cases, said that "the records and evidence show very clearly that the contempts of Judge Noyes and Frost were committed in pursuance of a corrupt conspiracy with Alexander McKenzie and with others not before the court, and therefore not necessary to be named, by which the properties involved in the suits mentioned in the opinion, among other properties, were to be wrongfully taken, under the forms of law, from the possession of those engaged in mining them, and the proceeds thereof appropriated by the conspirators." Comment is superfluous. Noyes was only sentenced to pay a fine; out of respect for his judicial position he was not sent to prison; he was successful in delaying the execution of his sentence; he and Frost continued to draw their salaries, even though Frost was in jail. The district attorney, Wood, who was sentenced to four months, also remained on the pay-roll of the Government, and it was not until February 1902, five months later, that their pay was stopped.

In the meantime, in September 1901, Judge Wickersham had been transferred from Juneau to Nome, and had begun to disentangle the maze of iniquity left by Noyes. But corruption was still rampant. The marshal, Richards, was packing juries, and the new district attorney, Griggsby, a Rough Rider appointed by President Roosevelt at the instance of the senator from North Dakota, was emulating his predecessor. Finally President Roosevelt caused a thorough investigation to be made, whereupon Richards and Griggsby were summarily removed.

It is to the honor of American journalism that the *Washington Post* was instrumental in exposing the whole shameful story, causing the Senate to call for an investigation. In the debate that followed, Stewart of Nevada made a clear statement of the whole amazing affair, and it was such a thorough exposure that, upon request, a large part of it was expunged from the Congressional Record. But it fulfilled its purpose. The new Attorney-General caused a search to be made for the

AN ESKIMO CAMP.

documents in the case against Noyes, and thus, 18 months after they were filed, the charges came to a hearing. The judge was dismissed on the records of his own court, which had lain in the Department of Justice for a year and a half.

It is fortunate that the conspirators were not more careful; if they had exercised greater caution it would have been even more difficult to circumvent them. McKenzie and Noyes did coarse work, they did not follow the forms of law, they did not cover their tracks, evidently expecting that their actions were not subject to review save at the hands of their allies at Washington. They failed to appreciate the fact that appeals from Alaska went to the Court of Appeals for the Ninth Judicial Circuit, sitting at San Francisco. This circuit included Alaska and the Hawaiian Islands. Moreover, McKenzie blundered in not realizing that this Court of Appeals could not be 'reached' by him or the like of him.

After Noyes had been punished, and after Wickersham had served temporarily, the appointment of Judge of the Second District was given to Alfred S. Moore, at the request of Quay, the Pennsylvania political boss and arch corrupter. Moore is judge now, and a poor judge too, from all I could learn while at Nome. He presides over a court infested by men most of whom, under the guise of lawyers, are mere tricksters and thieves. When the Third Beach was discovered, in the fall of 1905, the lawyers banded to make the most of their chance to get rich. Only one claim escaped litigation. The methods adopted varied: one of the most successful was to secure perjured testimony as to the marking of boundaries from men in the vicinity at the date of location, and thus shift the lines, moving one claim on top of the other, so as to cover the coveted ground. The buying of jurors was even easier, as the jury was not confined but allowed to go home; in some instances two detectives were employed in shadowing each juror to prevent either side from getting at him. Most of the cases were concocted by the lawyers, who received one third of the claim if retained by a miner and an exorbitant fee if retained by a company. Thus it became expedient, as a man said half in joke, half seriously, to deed a claim, after location, to a lawyer, then take a 75% 'lay', meaning, to work the ground on a 75% royalty, thus giving the lawyer 25%, because if the miner kept the claim and sunk a shaft to 'pay', the ground would certainly be 'jumped', and in the litigation that followed the owner would have to give the lawyer one third before he could begin to protect himself in a lawsuit. The lawyers have fattened on such practices. The looting of the mines by these methods was even farther reaching than the infamous conspiracy of 1900, and it has despoiled more individuals.

Unfortunately, the complexities of the mining law subserve the chicanery of lawyers whose life-work is to defeat justice and aid the purposes of tricky adventurers whose hope is to get rich without labor. Even today Nome is peculiarly cursed with corrupt lawyers; the leading practitioners of the town will accept a retainer from both sides in a lawsuit and while acting for a client they will take advantage of the confidence reposed in them to obtain knowledge concerning the ground and then buy an interest in property their client is likely to purchase later. The fees charged are exorbitant and the men who receive them are charged with inciting schemes leading to costly litigation and professional corruption. A weak judge presides. He is ignorant of mining law and unstable in his

WALRUSES ASLEEP ON THE ICE.

judicial opinions, so that his decisions are not respected. It is a disgrace to the American flag that justice should be such a by-word in Alaska, especially on the Seward Peninsula.

Every judge appointed in Alaska has sooner or later been fiercely attacked; while this has been due in part to the personal incapacity or corruption of the judge, inevitable under the system by which he is selected, yet even a good man would have a hard time, because the industries of Alaska are controlled largely by two or three strong corporations; if a judge gives a decision in favor of one of these, he is pretty sure to be abused by one or more of the others. Corporate influences dominate the nomination of the judges through the senators of the party in power, the naming of judges being a part of the spoil of political victory, so that while a President may be innocent of choosing an unfit man, he is helpless, it being the custom for him to distribute the patronage of his office in accordance with the wishes of the leaders of his party. Thus Carter of Montana and Quay of Pennsylvania, two notorious political bosses, were the senators who chose the judges whom McKinley appointed and whom even Roosevelt sent to Alaska. Until the judiciary ceases to be the footstool of a political spoils system, such scandals as the Noyes-McKenzie affair are possible, although in this case the development of crime was facilitated by the distance from the centre of government.

But these eruptions of greed and chicanery are not peculiar to any one country. Curiously enough, at the very time when McKenzie and Noyes were exploiting Nome, another group of spoofers were working Manchuria. Instead of Noyes was Alexieff, instead of McKenzie was Bezobrazoff, and in lieu of Chipps was Vonlalarski. The amiable and worthless character decorated with the name of Admiral was the tool of the two schemers, who in turn were supported and abetted by the Grand Dukes—the Grand Dukes of a hereditary bureaucracy playing the part of the Senators of an undeveloped democracy. Bezobrazoff and Vonlalarski got Alexieff to grant them concessions of various kinds, notably a timber concession on the Yalu; this, by causing trespass on the Korean side of the river, led to the Japanese war, which upset the scheme of the gang. In Nome there was no war, no bloodshed. One reason for such marvelously peaceable behavior under extreme provocation was the fact that the victims of the McKenzie-Noyes conspiracy were mostly Scandinavians unused to firearms or to other reprisals of an explosive character. As a matter of fact, 'gun play' was a minor feature of this burlesque of republican institutions, but the presence of the military served as an excellent deterrent. The officer in command conceived it to be his duty to abide by the directions of the Judge, who had authority from Washington. The national capital was the headquarters of this travesty upon representative government.

CHAPTER XXXIV.

THE RIDE TO OPHIR.

On August 30 a party consisting of C. H. Munro, T. M. Gibson, Jafet Lindeberg, A. E. Boyd, Scott Turner, and the writer left Nome for Ophir creek. Going aboard the *Flyer,* a 30-ton gasoline launch, we went eastward, keeping within sight of the coast as far as Solomon, where the river of that name enters Bering Sea, forming a shallow estuary. Landing would have been simple if we had not run aground and the voyage would have been pleasant if the cold spray had not been flavored with the smell of gasoline, but these are details in which any Alaskan traveler will recognize only bits of local color. At Solomon horses were awaiting us, and I found, much to my pleasure, that 'Tony' was allotted to me. Tony is the finest horse I have ridden in any of my travels about the world; spirited but gentle, strong and willing, with a steady trot that knows no weariness, I found myself mounted in a style not to be expected in a region where horses are scarce. The owner of Tony is T. M. Gibson; to his kindness I am deeply indebted. Henceforth I shall have two ideals of self-denying courtesy: The first is that of a woman, a musician and a singer, who plays perfectly an accompaniment for a friend, the friend singing badly a song which the accompanist sings exquisitely. The other is that of a man possessing a fine horse, most comfortably gaited, which he lends to a friend, while he himself accompanies the party on a hired animal.

Owing to sundry delays we did not leave Solomon until dark, intending to spend the night at the camp of the Three Friends, whither we had been bidden by W. L. Leland. The distance was only 7 miles, but we must have forded the river fully 17 times. As the horses had been standing in the stable all day and were cold, and as there were many of them together, stimulating to rivalry, we went helter-skelter over corduroy roads, moss, and gravel, taking the fords at a trot, if not at a gallop, splashing in and out of the river, plunging along in the dark, until I wondered which moment was to be my last. In an hour we saw a light ahead and found a man with a lantern to guide us to the road that soon ended at the door of Mr. Leland's hospitable abode. There, in the midst of the cold wilderness, we found "all the comforts of home." An abundant 'supper', a good smoke, and a most interesting discussion, were eventually followed by unfathomable sleep.

The next day opened wet and stormy. Encased in 'slickers', or water-proof coats, hats, and rubber hip-boots, we went to see the dredge of the Three Friends Mining Company. This successful enterprise was organized by three men who knew what they were about, so that the common blunders of the inexperienced were avoided. Firstly, the ground was carefully tested by drilling; secondly, a strong machine was designed on the model of one of the best dredges then working at Oroville, California; thirdly, an experienced crew was engaged; finally, care was taken to avoid frozen ground. These four precautions constitute the A B C of dredging in Alaska, but they are often disregarded. The Three Friends dredge is making handsome profits, and affords an example to the organizers of companies intending to dig for gold in river-bottoms.

The dredge has turned over the bed of the Solomon river, making successive piles of débris that look like the furrows of a giant plowman. Even a casual glance shows that this reject is not all gravel, but fragmentary rock. On most of the 70 acres that have been over-turned the bedrock is schist, but where the dredge was at work last August the gravel lay on limestone, into which the buckets bit courageously. Indeed, it was remarkable how this Bucyrus machine dug into the weathered limestone of the river-bottom, bringing up chunks of white rock that looked as if they had passed through a rock-breaker. It is astonishing what work a modern dredge will accomplish.

SOLOMON RIVER, ALASKA, SHOWING THE THREE FRIENDS AND THE NOME-MONTANA-NEW MEXICO DREDGES AT WORK.

Under the guidance of Mr. Leland we went on board the dredge and watched the operations, which, although familiar, are always interesting. Each bucket holds five cubic feet, the spacing of the buckets and the speed of their travel being such as to give the dredge an actual daily capacity of 3500 cubic yards. The gravel yields about 50 cents' worth of gold per yard, at a cost of 18 cents. As the gravel and two feet of the limestone on which it lay, were brought up by the buckets, they discharged upon a tabular shaking screen, 24 feet long and 15 feet wide; this screen-table is made of steel-plate punched with holes, through which the fine stuff, including the gold, is washed by the play of strong jets of water. The boulders and other refuse proceed to the rear of the dredge, where they are rejected, while the screened material is washed over a series of riffles, arranged in three tiers, to arrest the gold. Mr. Leland was kind enough to stop the dredge, so that by aid of the electric light we could see the gold nesting under the riffles—and it is always pleasant to see gold thus, not so much as the expression of wealth but as the evidence of skill in extracting it from places that seem a little beyond the reach of man.

In the afternoon the weather cleared sufficiently to warrant a start. As Tony was restless, I rode ahead of the party to the road-house on the East Fork of the Solomon river, obtaining directions from Mr. Leland. The trail was easy to follow, for it was mainly in the river. Soon after starting I found that I had to cross the river where the water had been deepened by a dam built by the dredging company; while this ford was not deep enough to compel the horse to swim it made me thoroughly wet, for I had declined the loan of rubber-boots as being awkward when riding. On the Seward Peninsula the rubber-boot is continually required; everything is wet; the moss is soaking, and even when on horseback the frequent fording of icy-cold streams renders such protection advisable. The roads, for the most part, cling to the river beds, where gravel affords fairly good footing as compared to the soggy tundra; in consequence, the road is in the river, and the river is in the road, making it immaterial whether it be labeled a water-way or a trail.

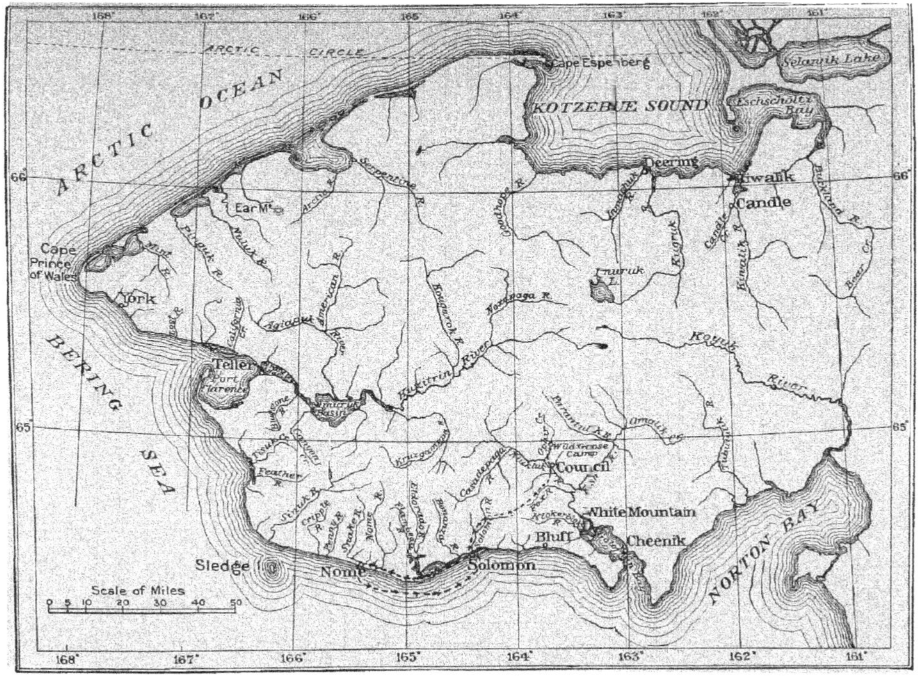

THE SEWARD PENINSULA.
Illustrating Route from Nome to Ophir.

A series of short trots over the bare gravel and a number of splashes through the stream brought me to a lateral gulch which I recognized, from description, as the one that led to the Big Hurrah mine. This is remarkable as the premier gold-quartz mine on the Peninsula, that is, a mine from which gold-bearing ore was extracted from a quartz lode, as distinguished from the gravel of the placer deposits constituting the source of nearly all the gold won in northern Alaska. The Big Hurrah was one of the undertakings of that fine old Californian, Charles D. Lane. The lode consists of several quartz veins traversing graphitic schist. The quartz is banded with graphite, which hindered amalgamation, although much of the gold was so free as to be visible. Many beautiful specimens came from this mine, and there is no doubt that the erosion of this lode was one of the factors enriching the Solomon river.

Proceeding up the river I saw several derelict dredges of rudimentary design and heard the coughing of the gasoline engine actuating one that was still at work. Tony pricked up his ears and gave signs of dissatisfaction with this asthmatic mechanism, but it was soon passed, like a memory of the crudities that have made the dredging branch of gold mining a nightmare of blunders.

The seven or eight miles were shortly covered; there had never been any danger of losing my way, for the river was a sufficient guide, and on the adjoining hillslope the line of the Council City & Solomon River Railway afforded assurance of the direction to be taken, although it provided no more than that, being only the sign of a dormant system of transportation. No trains puffed along the quiet valley, no whistle broke the heavy silence; the railway was not in operation, because the traffic was inadequate. However, there it was: a thread linking the desolate foreground to a radiant distance, connecting this great lone land with home. I liked to see it.

Meanwhile Tony maintained a steady trot and when the trail passed under the railway bridge I found myself close to a group of buildings which I knew must be the road-house. So it proved. Getting rid of my wet clothes and borrowing some commodious slippers, I was soon toasting beside the hot stove,

COUNCIL. A PIONEER SETTLEMENT.

while waiting for my friends, who arrived half an hour later. We fared well that night. Of course, road-houses vary. Fortunately, I did not sample many of them while in Alaska. Sheets are not provided; instead, the traveler wraps himself in blankets used by his predecessors, and it is only the fatigue following exercise in a cold bracing air that enables him to overcome the anticipation of attack from various forms of insect life. In the morning the traveler is invited to "take a shot" before breakfast; this pre-prandial drink being offered by the proprietor as a courtesy. It is also customary for the bar-tender to invite the habitues of the establishment, or those who have patronized the place on the preceding night, to take a drink before breakfast at the expence of "the house." It is a villainous custom and creates a vicious habit. Men get so used to it that they cannot eat their breakfast without the alcoholic stimulus, or, as Dr. Wiley would phrase it, surprising the expanse of expectant pepsin with a flood of ardent alcohol. We needed no such stimulant. We felt like eupeptic heroes. The food served was excellent, for this road-house was managed by a clean thrifty woman. The raspberries from the neighboring moor and the fish from the river were both delicious. Even if the eggs were overdue that did not matter, for we were used to 'case' eggs; in fact, one of my friends at Dawson told me that he had become so accustomed to the ripe rich flavor of case eggs that when he went home and his mother took pains to have perfectly fresh eggs served at breakfast on the morning following his arrival, he complained that they were insipid! Eggs intended for mining camps and other distant places are packed in sectionalized cardboard boxes, each egg being in a separate partition, and 30 dozen in one 'case.' In order to preserve such eggs they undergo treatment previous to being packed, the essential feature of all the various methods being exclusion of air. In the old days eggs were buried in salt and kept in a cool place. Nowadays eggs intended for transport are immersed in a solution of water-glass or in lime-water; the first of these is a soluble silicate of soda, while the second is ordinary unslaked lime in water. By this treatment the decay is retarded and in a cold country they keep for a long time; the freshest egg eaten at Nome must have been laid five months, while the age of the oldest a polite man would no more care to guess that the years of an old maid. Even the eggs that seem but unhatched chickens are palatable to hungry men. Another staple article of food is condensed milk; as to that it is safe to say that good condensed milk is more wholesome than most of the milk sold in cities; certainly, the miner gets to like it, especially with his coffee, and he adds it to the wild raspberries and blueberries with gusto. Habit is second nature.

Leaving the East Fork next morning we crossed the divide separating the watersheds of the Solomon and Fox rivers. At noon we halted at Hatch's cabin on I. X. L. creek. Etiquette in Alaska, or elsewhere, requires that if you use another man's cooking utensils, you shall clean them after using. Also you shall put at least as much firewood beside the stove as that consumed by your own use. For the rest, the prospector in the North is generous, and in his absence, is not averse to the consumption of his bacon and beans, leaving it to you to return the hospitality when he comes to your 'creek.'

The scenery in this part of the Peninsula is subdued: Low rounded ridges, clad in tundra, are crested with outcrops of schist, of fantastic form, due to accidents of weathering on a rock of tabular structure. As seen through flying veils of mist these isolated rocks took the shape of men and monsters; to the first invaders of the wilderness they must have been uncanny, and to an imagination distorted by hunger or weariness they probably loomed like the figures of a distempered dream. To those who know Otago, in the South Island of New Zealand, I can say that in rock-weathering this part of the Seward Peninsula resembles the ridges near Barewood.

The landscape was not without color, for the first frosts had already touched the foliage; in the valleys the willows and blueberry bushes had the glint of gold, and on the hillsides the imperial purple of autumn was thrown over a wilderness whose furthest rim was silhouetted against a gray sky.

Tony was full of vigor and led the cavalcade. In the stream, repeatedly forded, the trout darted from underneath the hoofs of the horse, while the frightened duck flew in near circles. Finally, leaving Fox creek I ascended a short rise and found myself on the edge of the Niukluk valley. A perfectly straight road, about five miles long, stretched across the tundra to Council, a cluster of white buildings on the farther bank of the river. This bit of good road was built by the Road Commission and consists of corduroy, covered with moss and topped with gravel. It connects the trail or water route of the Niukluk with the Solomon river, and has proved a great help in the haulage of supplies to the mines.

Council saw the beginning of gold mining on the Seward Peninsula, as has already been recorded; but it is quiet now, with a population of 200 only. At the ford of the Niukluk the members of our party re-united and after plunging through the river we skirted the village, taking a trail that followed the left 'limit' of the Niukluk. A curious bit of physiography is here exhibited: the Niukluk and a tributary stream—Ophir creek—are parallel, occupying the same valley without loss of identity; thus for a mile Ophir creek runs within 100 yards of the Niukluk river. In order to float a dredge from Council to Ophir, a cut was blasted between the two water-channels. Turning into the canyon of Ophir creek we passed the dredge, called the Blue Goose, and just above it we reached our destination, the camp of the Wild Goose Mining & Trading Company, distant 42 miles from Solomon.

Here we spent four days. There was much to see in the way of mining operations, including the use of elevators and dredges, but this is not the place for technical details. They will be found elsewhere.* During the daytime we rode or walked to the various centres of mining activity, as, for instance, to the Blue Goose dredge. This we found idle, owing to an accident such as may befall even the best-managed enterprise. An unexpected encounter with a patch of frozen ground had strained the bucket-line, breaking some of the teeth of the main driving wheel, and some of these broken pieces becoming caught in the pinion had cracked the rim of the driving-wheel

*_Mining and Scientific Press_, November 28, 1908; also March 13, 1909.

A TYPICAL LANDSCAPE ON THE SEWARD PENINSULA.

itself. At the same time the big wooden anchorage or 'spud' was shattered. Altogether it looked like a serious disaster, for Ophir creek is a long way from timber-yards and foundries. Nevertheless, ingenuity and hard work overcame the damage without delay; anxious to make the most of the only too brief season, the crew worked by night and day, with a right good will, so that within three days the machine once more was digging. G. H. Russell, the manager, and Jafet Lindeberg, the president of the company, are to be congratulated on this example of efficiency. As stated already, the accident was due to 'bucking' frozen ground; in fact, if anything goes wrong with a dredge in the North, *cherchez la glace*!

This little dredge is a self-contained machine; on board I found a machine-shop, smithy, and mess-room. The crew take their midday meal on board, and as the soup is agitated by the vibrations of the dredge they must feel as if they were on a Yukon steamer aground. The running time is 110 days each summer, and of this 31 per cent is lost by stops for repairs and other causes. About 100,000 cubic yards are dug, for a yield of $85,000 and a profit of $50,000. This is encouraging.

On September 3 we rode to No. 24 Above, or 8 miles beyond the camp on Discovery where the Wild Goose company has its headquarters. We passed the old camp on No. 15 Above where Charles Lane lived several years ago. Claim No. 15 yielded $1,400,000. On No. 14 a piece of ground 1100 feet long and 125 feet wide yielded $800,000. At the upper end of No. 20 the gravel was worth $20 per cubic yard. It was well named Ophir.

Yet, although so near to Solomon, these gold diggings are not imputed as the source of the Jewish king's wealth. In the Zambesi region of central Africa there is a mountain named Fura, a name taken by the Portuguese from Arab traders, who corrupted it from the Hebrew word Ophir. Indeed, the word Africa itself, which the Romans took from the Carthaginians, is a modification of Afur or Aufur; thus from Ophir, through Afir, Afer, Afur, was Africa evolved. Carl Peters is my authority, but he went farther, he went to the Zambesi. Between that great river and the Sabi, in modern Rhodesia, he found many thousand old mines some of which, ten years ago, afforded the foundation for company finance in London. These old workings are scattered over a region covering 750,000 square miles, and within this area are ruins of cities, fortresses, and temples bearing evidence of the ancient worship of Baal-Ashera and of the civilization that characterized the Himyarites of southern Arabia. From the hot sands beside the Red Sea to the frozen gravel by Bering Sea is a far cry, but the human motive is the same; despite changes of time and place, the digging of gold has been the spur to exploration and the pioneer of civilization.

At claim No. 19 on Ophir a sharp turn in the creek coincides with a sharp turn in the sentiment of these pages, for here is recalled the long litigation known technically as Walton *v.* Wild Goose. Walton was, and is, a colored lady, whom I happened to see at Nome, while at the Golden Gate hotel. A telephone in the lobby was used by Miss Walton with such noisy insistence as to give an idea of her possibilities as a litigant. In local history she will figure as the poor lone negro-woman who sued a soulless corporation, and lost. It was thus: In the early days of Ophir creek the prospectors used printed location forms claiming the bed of the creek and "the meanderings thereof." On No. 19 the stakes were not placed in accordance with this description, but in a straight line across the bend of the channel. A claim-jumper took advantage of this fact, and Miss Walton became a partner with him. After a long struggle in the courts it was decided, as is usual in such cases, that the actual position of the stakes took precedence over the description appearing on the location notice. This decision was appealed and sustained. Another interesting point arising in this litigation was the assertion that Capt. Walker, the locator by proxy, had not done sufficient assessment work. Walker was stationed at St. Michael; after the close of navigation he sent three men overland, instructing them to comply with the mining regulations. They dug a trench, which, it was said, took 10 days of work. The Walton people claimed that such a trench could be dug in 3 days, and they caused such a trench to be dug in that amount of time. The Judge, however, instructed the jury that they might consider the distance from the source of supplies, and the consequent expense involved, in estimating the cost of the work done on the claim. It was 175 miles to St. Michael and cost the captain $300 to send the three men, therefore the jury found that he had shown good faith in maintaining his title by doing the work required by law.

On 24 Above we lunched with J. L. Wilson and saw the operation of his bucket dredge, a small machine moving on a turn-table; this dredge was doing good work, digging to a depth of 15 feet, with 12 to 15 horse-power, obtained from the combustion of distillate.

That evening snow fell and the rapidly shortening days gave warning to the mine operators that their season was already waning. The absence of smoke, either from forest fires or factories, caused the atmosphere to be wonderfully clear. It was to this cause that I impute the marvelous coloring at sunset: A gray sky, belted near the horizon by a broad band of pink; under it purple clouds, against which the long curve of the hills and the sweeping contours of the darkly green tundra were thrown in exquisite purity of line and color.

On the morning of September 4 we started on the return journey, leaving Ophir creek at 9:40 a.m. and reaching Leland's camp at 5:10 p.m., thus covering the 42 miles in 7½ hours, including the half-hour halt at midday in Skookum gulch. It was cold and windy weather, culminating in a slight fall of snow. On this day my saddle-bags contained 561 ounces of gold, in two bars, but the additional weight did not bother Tony, who maintained a steady trot over the uneven trail. Crossing the divide into the valley of the Solomon, I noted the dome-shaped hills of limestone, weathered and gray, with tundra clinging to their feet. It was a forbidding landscape: cold gray hills against a wintry sky. But the air was bracing and the swinging gait of the horse made exhilarating exercise, ending in a comfortable rendezvous at Oro Fino, as the camp of the Three Friends is called. At 7:40 the next morning we were in the saddle again, arriving at Solomon at 8:25; here the gold was taken from me for shipment by the *Flyer*, that unhappy vessel being visible off-shore, where she was again stranded on a bar; as the tide was ebbing, the prospects of a

BERING SEA.

start were poor. This was fortunate, for it decided us to ride to Nome on our trusty horses rather than risk a voyage along the coast in a miserable launch. Munro, Turner, and I left Solomon at 9:15, and then commenced the finest ride of my life.

The road ran along a sand-spit, formerly a bar, but now green with tough grass. With firm ground underfoot and willing horses, the telephone poles flanking the road in long perspective were passed in quick succession. Only a hundred yards to the left was Bering Sea, smooth and sunlit, with no suggestion of the storms that had flung the gray driftwood high on the edge of the land. On the right, also a hundred yards away, a long lagoon mirrored the blue and white of the sky; inland stretched the tundra, in the tawny livery of autumn, crossed by the purple shadows of traveling clouds, while in the distance northward rose the limestone hills whence come the Bonanza and Flambeau rivers. At 18 miles from Solomon the road is interrupted by an estuary at the mouth of which is Port Safety. Here is the Government wireless telegraph station. As we waited for the ferry the only sound to be heard was the coughing of the gasoline engine that runs the dynamo of the wireless equipment. This station is said to be the most successful of its kind in catching messages from passing ships, and it is also the only means of telegraphic communication between the Seward Peninsula and the outer world, messages being transmitted to St. Michael, whence they are forwarded overland by the military telegraph system, which crosses Alaska by way of Eagle and Fairbanks.

Beyond Port Safety the road continued excellent, clinging to the edge of the land, so that we were always close to the sea. Several Eskimo camps were passed. Countless dogs made the quiet air vibrate with their melancholy howls. The squaws were picking blueberries, using wooden rakes made from the covers of cigar-boxes. One young woman in a gay parka was attended by a young man who lay idly watching his belle at her pleasant task. An Eskimo idyll—the same old game! On a tent-pole the skin of a hair-seal was drying and on other frames fish were hanging like bananas. We saw an *oomiak* or

AFTER THE RIDE TO OPHIR. SEPTEMBER 1908.

T. W. Gibson. Scott Turner. T. A. Rickard. C. H. Munro. Jafet Lindeberg.

native boat containing a family of ten, the father steering with the paddle, while the boat was towed by six dogs, tended by a young man and woman, walking together. At 12:45 we reached the road-house near Cape Nome, remaining there until 2:30, so as to give the horses a proper rest. It was from Cape Nome that the district obtained its name; on the western side of the cape we abandoned the road and took to the beach, riding along the shore for 10 miles, as far as Fort Davis.

The tide is out and the firm wet sand affords excellent footing. Our horses are willing and full of life, emulous to keep ahead, so that they move gladly. Tony will not be denied; maintaining a fast trot as if he had just come out of his stable. The cold wind blows from Bering Sea, the breakers curl to the shore, the horse treads on the silver fringe of the waves as they ripple over the sandy beach; the strong pulsating movement of the noble animal, the exhilaration of an air fresh as the breath of dawn, the feeling of perfect freedom and unlimited expansiveness, a sense of complete well-being and gay adventure, touch every tingling nerve and stimulate every willing muscle, until I shout with delight. It is well for a brief space to feel like a viking when the world was young, as if the complexities of civilization had been swept away by the cold breath of the sea, as if money and business, steam and electricity, factories and libraries, no longer existed, while the tide of life fills every corner with exultant joy.

At Fort Davis we had to return to the road, for the beach was being mined for gold, but it was only three miles to Nome. Although we had ridden 43 miles in 5½ hours of actual travel our horses were still unwearied, which is further evidence of the invigorating nature of this Northern atmosphere. We rattled down the planked way of the long Front street, and dismounted at the Wild Goose office at 4:30. After a change and a bath we made tracks for the Royal Café and there celebrated the termination of a glorious trip by dining on wild goose and reindeer stew, washed down with a bottle of Californian burgundy, followed by the smoke that makes all men brothers.

CHAPTER XXXV.

SAN FRANCISCO.

The journey from Nome to Seattle calls for no special comment. The *Northwestern,* of the Alaska Steamship Company, was not as big as an Atlantic liner, having a register of 3396 tons and a length of 350 feet only, but she was comfortable. Her hold being almost empty, she threshed about in the trough of the seas, but after the river boats it seemed luxurious to be on a well appointed vessel. The passengers included a number of mining men, among whom were several who had taken part in the early development of Nome and whom therefore I found particularly interesting. Two days after leaving Nome a light on the port side indicated the beacon guarding the strait of Unimak, in the Aleutian islands. No stop was made at Dutch Harbor, we passed through the strait and our course was laid due east, emphasizing how far "the path of Empire" has led westward. The best day's run was 316 knots; the poorest, 284. On September 15, seven days out, we sighted Vancouver island, and the next sunrise saw us in the quiet waters of Puget Sound. It was a fine morning; to starboard the snow-clad peaks of the Olympic range, which three months earlier, on the outset of my journey to the Northwest, had seemed a bit of grand scenery, now wore a familiar look and spoke of the nearness to home. Landing at Seattle, I found that the steamer for San Francisco did not leave for 48 hours. Accordingly, after spending a night in Seattle, I took the boat for Victoria. When the *Governor,* of the Pacific Coast Steamship line, arrived on the evening of September 18 I went aboard and on September 21 the long journey ended at the port of San Francisco.

At break of day the *Governor* was gliding slowly over a gray sea into a silver mist from which issued the hoarse warnings of foghorns. The opalescent light illumined the inflowing tide and glanced upon the sails of a fishing boat; the delicious stillness was only broken by the lapping of the water against the sides of the ship, until suddenly the fierce blast of our siren made the air vibrate with waves of sound that were quickly swallowed by the fog.

It was early dawn; as the mist took form momentarily the blue sky lit by the morning sun flashed a Californian greeting overhead. Thereupon the white curtains of the fog were swept aside, and like a splendid scene in a theatre, the whole panorama of the Bay of San Francisco was spread to view. We were close to Alcatraz, the island prison, which like a fortress bars the Golden Gate; to the left were the hills of the Marin shore, with Sausalito, like Sorrento, clinging to the cliffs at whose feet the white sails come and go with the sea-gulls; ahead rose Angel island, with suggestions of naval activity, and Goat island, the Capri of these waters; across the calm expanse of the Bay, now laughing in the sun, stood the Contra Costa foot-hills, with all the familiar lines of grassy slope and oaken glade, at the base of which the linked communities of Berkeley, Oakland, and Alameda lay spread in peaceful security, forgetful of the tremor that had rocked the earth on the morning of April 18, 1906. The white buildings of the University, framed in groves of eucalyptus and acacia, were distinguishable, and over the intervening space of water the ferry-boats churned in orderly succession to and from the wharves of San Francisco. She, the warder of two continents, whom the poet in prophetic words called "indifferent of fate," rose above the swift tide and filled the picture, as she fulfilled my imagination. I saluted the most imperturbable of cities, the survivor of earthquake and fire, the vice-ridden, graft-cursed community, that, with heroic courage and splendid audacity had risen in self-assertive strength from ruins scarcely cold. Two years ago she lifted an ashen face to the blue of a pitilessly perfect sky; now the same site was covered by splendid structures of steel and stone, traversed by orderly streets, and thronged by happy faces.

On August 1, 1836, Jacob P. Leese wrote from Yerba Buena to a friend at Monterey, California, saying: "Leigh low and look out sharp for this Place as we are a gowing to do a snorting Buisiness. Do not make yourself uneasy about Buisiness here for all gows on straight." The Yerba Buena thus commended was the settlement that became San Francisco, and Jacob Leese was the first merchant of what was destined to be a great mercantile community. His optimism, told in homely language and early reformed spelling, has often been expressed since then in more resplendent phrase; it has passed from words to deeds. The barren sand-hills facing the surf of the Pacific have been made gracious with verdure; on the wind-swept dunes rise many-storied structures, knit with steel and clothed with stone; the shallows of the shore have been won to traffic, and the waters of the estuary bear the commerce of a continent. Yerba Buena has emerged from its Spanish background and has become the San Francisco of an American California.

San Francisco: The name has a melancholy cadence to those who have known her in the days of her prosperity, her disaster, and her shame. Today she guards the portal of the Pacific and takes it upon her to forbid the entrance of the coolies of Japan, with a manner in which are blended dignity and burlesque; in the midst of a reconstruction that challenged the admiration of the world she sent her indicted mayor, the first violin of a vaudeville, and the latest expression of labor unrest, to represent her in an international negotiation at Washington; in the very act of rising gloriously from the effects of a conflagration that enkindled the generosity of the world, she exposed a scandal of municipal corruption that made the smoke of her fire drift to heaven like the reek of a tannery. San Francisco epitomizes the greatness and the smallness of humanity; those who love her know that beneath the tawdry politics and the reckless industrialism beats the warm heart of a great personality. For cities have characteristics that give them an identity. The City beside the Golden Gate wears her heart upon her sleeve, she is the most Bohemian of her kind, neither self-conscious nor conventional, artless as a spoilt child and sensitive as a woman. Proud of her origin as the haven of the Argonauts, confident in the promise of her dominion of the Pacific, headstrong in the insistence upon her supposed rights, serene amid the shock of cosmic unrest, she appeals to the imagination of the poet and to the perception of the artist. It was well to have known her at her worst, under her mountebank mayor and thieving political boss, in the thrall of municipal corruption and amid the stench of putrifying incivism; it was well to have known her at her best, facing a catastrophe that looked like the end of the world and repairing a disaster that had no promise of a morrow; it was well to have shared her ruin and her resurrection, to have felt her shame and gloried in her greatness.

As the tattered veil of fog was blown southward by the morning breeze, I looked at the awakening city, now radiant in the sunlight, and seemed to see the day when San Francisco should glow in the dawn of a civic regeneration and feel the bracing air of a sane patriotism, before which the greed and graft, the corruption and chicanery, should be blown as a fog from the sea. Then, firmly poised on her peninsula, San Francisco shall face her destiny with proud serenity. With her back to the trails that brought her people over mountain and plain from the distant seaboard on the other side of the continent, she shall face the sunset with an earnest face. The Atlantic breakers sound faint; the surf at her feet thunders a welcome. Her eyes have caught the glow of the setting sun and in her ears is the song of the Pacific, bringing the promise of commerce with the immemorial East that has become a new West.

BOOKS OF REFERENCE

'Through the Yukon Gold Diggings'. By J. E. Spurr. 1900.

'Three Years in the Klondike'. By Jeremiah Lynch. 1904.

'Alaska and the Klondike'. By Angelo Heilprin. 1903.

'The Geography and Geology of Alaska'. By Alfred H. Brooks. 1906.

'Glacier Bay and Its Glaciers'. By Harry Fielding Reid. U. S. Geological Survey. Sixteenth Annual Report, Part I.

'The Looting of Alaska'. By Rex E. Beach. Appleton's Magazine. January to May, 1906.

'Voyage of Discovery'. By George Vancouver. 1798.

'History of California'. By Theodore H. Hittell. 1898.

'The Land of Nome'. By Laurier McKee. 1902.

'Alaska, The Great Country'. By Ella Higginson. 1908.

'Alaska and Its Resources'. By William H. Dall. 1897.

'Alaska.' Vol. I, II, and III. Harriman Expedition. 1902.

'The Gold Placers of Parts of Seward Peninsula, Alaska'. Bulletin No. 328, U. S. Geological Survey.

'The Ice Age in North America'. By G. Frederick Wright. 1888.

'History of Alaska'. By H. H. Bancroft.

'Guide-book to Alaska'. By Eliza R. Scidmore. 1899.

INDEX

Spotted something that isn't quite right?

Let us know and we'll fix it for future reprints. And as a thank you for your diligence, we'll send you a free Sothis book of your choice.

publisher@sothis.press

This page intentionally left blank

Except of course for that line above. And this one.
And whilst we have your attention, perhaps you'd
like to follow us on Instagram @sothis.press. Go on.

Made in the USA
Monee, IL
08 November 2026

68106871R00239